Methods in Molecular Biology

Series Editor
John M. Walker
School of Life and Medical Sciences
University of Hertfordshire
Hatfield, Hertfordshire, UK

For further volumes:
http://www.springer.com/series/7651

For over 35 years, biological scientists have come to rely on the research protocols and methodologies in the critically acclaimed *Methods in Molecular Biology* series. The series was the first to introduce the step-by-step protocols approach that has become the standard in all biomedical protocol publishing. Each protocol is provided in readily-reproducible step-by-step fashion, opening with an introductory overview, a list of the materials and reagents needed to complete the experiment, and followed by a detailed procedure that is supported with a helpful notes section offering tips and tricks of the trade as well as troubleshooting advice. These hallmark features were introduced by series editor Dr. John Walker and constitute the key ingredient in each and every volume of the *Methods in Molecular Biology* series. Tested and trusted, comprehensive and reliable, all protocols from the series are indexed in PubMed.

Imaging Cell Signaling

Edited by

Christoph Wülfing

School of Cellular and Molecular Medicine, University of Bristol, Bristol, UK

Robert F. Murphy

Computational Biology Department, Carnegie Mellon University, Pittsburgh, PA, USA

Humana Press

Editors
Christoph Wülfing
School of Cellular and Molecular
Medicine
University of Bristol
Bristol, UK

Robert F. Murphy
Computational Biology Department
Carnegie Mellon University
Pittsburgh, PA, USA

ISSN 1064-3745 ISSN 1940-6029 (electronic)
Methods in Molecular Biology
ISBN 978-1-0716-3833-0 ISBN 978-1-0716-3834-7 (eBook)
https://doi.org/10.1007/978-1-0716-3834-7

This Humana imprint is published by the registered company Springer Science+Business Media, LLC, part of Springer Nature.
The registered company address is: 1 New York Plaza, New York, NY 10004, U.S.A.

Paper in this product is recyclable.

Preface

The elucidation of mechanisms of signal transduction is critical for understanding the regulation of cell function and underpins much of drug development. Key to signal transduction are molecular interactions between proteins, lipids, and small molecules and biochemical modifications thereof. It is increasingly appreciated that, inside live cells, the efficiency of such interactions is governed by the localization of the signaling intermediates. Significant progress in methods to image cells over the last few years is generating greatly improved access to the subcellular organization of signaling, enabling hitherto inaccessible research in the physiological regulation of cell function and its therapeutic exploitation. There are a number of key challenges in the imaging of signal transduction. Cells need to be isolated such that they retain key elements of their in vivo function on the microscope stage. The imaging of signal transduction needs to be extended from the conventional diffraction-limited visualization of large clusters of individual signaling intermediates to single molecule resolution and the parallel imaging of multiple signaling intermediates. Computational methods are then required to analyze the resulting large and complex datasets and build models to predict the impact of manipulations and to establish causality. This can be matched by spatiotemporal manipulation to test imaging-generated predictions. The imaging of each signal transduction system requires the generation of a bespoke imaging strategy. In this volume, we have collected chapters that illustrate how each of the steps of a comprehensive imaging experiment, cell preparation, imaging itself, cellular manipulation, and computational analysis has been addressed in various ways. We hope that together these chapters will provide inspiration and guidance in the effective tailored design of projects to image, model, and manipulate signal transduction.

Bristol, UK — *Christoph Wülfing*
Pittsburgh, PA, USA — *Robert F. Murphy*

Contents

Contributors

GÖKTUĞ ABA • *Department of Cell and Chemical Biology, Leiden University Medical Center, Leiden, The Netherlands*

MORTEZA ARAMESH • *Division of Biomedical Engineering, Department of Materials Science and Engineering, Uppsala University, Uppsala, Sweden; AM4Life Comptenence Centre in Additive Manufacturing for the Life Sciences, Uppsala University, Uppsala, Sweden; Department of Information Technology and Electrical Engineering, Institute for Biomedical Engineering, ETH Zurich, Zurich, Switzerland*

NICHOLAS ARIOTTI • *Institute for Molecular Bioscience, Centre for Cell Biology of Chronic Disease, The University of Queensland, St. Lucia, Brisbane, Australia*

GABRIEL AU • *Department of Biology, Krieger School of Arts and Sciences, Johns Hopkins University, Baltimore, MD, USA*

DULAMKHUU BUJAA • *Physique et Mécanique des Milieux Hétérogènes, ESPCI Paris, PSL University, CNRS, Univ Paris, Sorbonne Université, Paris, France*

ANCHI S. CHANN • *Optical Science Centre, Faculty of Science, Engineering & Technology, Swinburne University of Technology, Hawthorn, Australia; Immune Signaling Laboratory, Peter MacCallum Cancer Centre, Melbourne, VIC, Australia; Department of Immunology, Mayo Clinic, Scottsdale, AZ, USA*

MIRREN CHARNLEY • *Optical Science Centre, Faculty of Science, Engineering and Technology, Swinburne University of Technology, Hawthorn, Australia; Immune Signaling Laboratory, Peter MacCallum Cancer Centre, Melbourne, VIC, Australia*

WEI-YU CHI • *Department of Pathology, Johns Hopkins Medical Institutions, Baltimore, MD, USA; Weill Cornell Medicine, New York, NY, USA*

COLIN J. CHU • *UCL Institute of Ophthalmology, London, UK; National Institute for Health and Care Research Biomedical Research Centre at Moorfields Eye Hospital and UCL Institute of Ophthalmology, London, UK*

DAMIEN CUVELIER • *UMR 144, Institut Curie and Institut Pierre Gilles de Gennes, PSL University, CNRS, Paris, France*

VAIBHAV DHYANI • *Bioimaging and Data Analysis Lab, Department of Chemical Engineering, Indian Institute of Technology Hyderabad, Sangareddy, Telangana, India; Optical Science Centre, Faculty of Science, Engineering & Technology, Swinburne University of Technology, Hawthorn, Australia*

OLIVIA DU ROURE • *Physique et Mécanique des Milieux Hétérogènes, ESPCI Paris, PSL University, CNRS, Univ Paris, Sorbonne Université, Paris, France*

MUNA FUYAL • *Division of Biomedical Sciences, Warwick Medical School, University of Warwick, Coventry, UK*

LOPAMUDRA GIRI • *Bioimaging and Data Analysis Lab, Department of Chemical Engineering, Indian Institute of Technology Hyderabad, Sangareddy, Telangana, India*

JANETT GÖHRING • *Institute for Hygiene and Applied Immunology, Center for Pathophysiology, Infectiology and Immunology, Medical University of Vienna, Wien, Austria*

TRESSAN GRANT • *Department of Life Sciences, University of Bath, Bath, UK; School of Cellular and Molecular Medicine, University of Bristol, Bristol, UK*

Julien Heuvingh • *Physique et Mécanique des Milieux Hétérogènes, ESPCI Paris, PSL University, CNRS, Univ Paris, Sorbonne Université, Paris, France*
Cheng-Chieh Huang • *Department of Pathology, Johns Hopkins Medical Institutions, Baltimore, MD, USA*
Chuan-Hsiang Huang • *Department of Pathology, Johns Hopkins Medical Institutions, Baltimore, MD, USA; Department of Cell Biology, School of Medicine, Johns Hopkins University, Baltimore, MD, USA; Center for Cell Dynamics, School of Medicine, Johns Hopkins University, Baltimore, MD, USA*
Johannes B. Huppa • *Institute for Hygiene and Applied Immunology, Center for Pathophysiology, Infectiology and Immunology, Medical University of Vienna, Wien, Austria*
John R. James • *Division of Biomedical Sciences, Warwick Medical School, University of Warwick, Coventry, UK*
Anumita Jawahar • *Physique et Mécanique des Milieux Hétérogènes, ESPCI Paris, PSL University, CNRS, Univ Paris, Sorbonne Université, Paris, France; UMR 144, Institut Curie and Institut Pierre Gilles de Gennes, PSL University, CNRS, Paris, France*
Ralf Jungmann • *Faculty of Physics, Center for NanoScience, Ludwig Maximilian University, Munich, Germany; Max Planck Institute of Biochemistry, Planegg, Germany*
Ute Jungwirth • *Department of Life Sciences, University of Bath, Bath, UK; Cancer Research Horizons Therapeutic Innovation, Newcastle Drug Discovery Group, Translational and Clinical Research Institute, Newcastle University, Newcastle, UK*
Tua Karling • *School of Biochemistry, University of Bristol, Bristol, UK*
Florian Kellner • *Institute for Hygiene and Applied Immunology, Center for Pathophysiology, Infectiology and Immunology, Medical University of Vienna, Wien, Austria*
Valentin Laplaud • *LadHyX, CNRS, Ecole polytechnique, Institut Polytechnique de Paris, Palaiseau, France*
Emily Lay • *Department of Life Sciences, University of Bath, Bath, UK*
Changtai Li • *UCL Institute of Ophthalmology, London, UK*
Alex X. Lu • *Microsoft Research New England, Cambridge, MA, USA*
Elina Mäntylä • *BioMediTech, Faculty of Medicine and Health Technology, Tampere University, Tampere, Finland*
Yuxuan Meng • *UCL Institute of Ophthalmology, London, UK*
Alan M. Moses • *Department of Cell & Systems Biology, University of Toronto, Toronto, ON, Canada*
Robert F. Murphy • *Computational Biology Department, Carnegie Mellon University, Pittsburgh, PA, USA*
Gregory S. Patient • *UCL Institute of Ophthalmology, London, UK*
Cecilia Persson • *Division of Biomedical Engineering, Department of Materials Science and Engineering, Uppsala University, Uppsala, Sweden; AM4Life Competence Centre in Additive Manufacturing for the Life Sciences, Uppsala University, Uppsala, Sweden*
Matthieu Piel • *UMR 144, Institut Curie and Institut Pierre Gilles de Gennes, PSL University, CNRS, Paris, France*
Sarah M. Russell • *Optical Science Centre, Faculty of Science, Engineering and Technology, Swinburne University of Technology, Hawthorn, Australia; Immune Signaling Laboratory, Peter MacCallum Cancer Centre, Melbourne, VIC, Australia; Sir Peter MacCallum Department of Oncology, The University of Melbourne, Melbourne, VIC, Australia*

FERENC A. SCHEEREN • *Department of Cell and Chemical Biology, Leiden University Medical Center, Leiden, The Netherlands*
LUKAS SCHRANGL • *Institute of Applied Physics, TU Wien, Wien, Austria*
FLORIAN SCHUEDER • *Department of Cell Biology, Yale School of Medicine, New Haven, CT, USA; Department of Microbial Pathogenesis, Yale School of Medicine, New Haven, CT, USA*
GERHARD J. SCHÜTZ • *Institute of Applied Physics, TU Wien, Wien, Austria*
THOMAS H. SHARP • *Department of Cell and Chemical Biology, Leiden University Medical Center, Leiden, The Netherlands; School of Biochemistry, University of Bristol, Bristol, UK*
HUANGQINGBO SUN • *Computational Biology Department, School of Computer Science, Carnegie Mellon University, Pittsburgh, PA, USA*
YUAN TIAN • *UCL Institute of Ophthalmology, London, UK*
PAUL VERKADE • *School of Biochemistry, University of Bristol, Bristol, UK*
JOSEPH VERMEIL • *Physique et Mécanique des Milieux Hétérogènes, ESPCI Paris, PSL University, CNRS, Univ Paris, Sorbonne Université, Paris, France; UMR 144, Institut Curie and Institut Pierre Gilles de Gennes, PSL University, CNRS, Paris, France*
GERNOT WALKO • *Department of Life Sciences, University of Bath, Bath, UK; Barts Centre for Squamous Cancer, Institute of Dentistry, Queen Mary University of London, London, UK; Centre for Oral Immunobiology and Regenerative Medicine, Institute of Dentistry, Queen Mary University of London, London, UK*
SUYANG WANG • *Department of Pathology, Johns Hopkins Medical Institutions, Baltimore, MD, USA*
HELEN WEAVERS • *School of Biochemistry, University of Bristol, Bristol, UK*
CHENTAO WEN • *RIKEN Center for Biodynamic Research, Kobe, Japan*
JR-MING YANG • *Department of Pathology, Johns Hopkins Medical Institutions, Baltimore, MD, USA*

Chapter 1

Rapid Preparation of Living *Drosophila* Pupal Macrophages for Ex Vivo Imaging

Tua Karling and Helen Weavers

Abstract

The fruit fly *Drosophila* is a well-established invertebrate model that enables in vivo imaging of innate immune cell (e.g., macrophage) migration and signaling at high spatiotemporal resolution within the intact, living animal. While optimized methods already exist to enable flow cytometry-based macrophage isolation from *Drosophila* at various stages of development, there remains a need for more rapid and gentle methods to isolate living macrophages for downstream ex vivo applications. Here, we describe techniques for rapid and direct isolation of living macrophages from mature *Drosophila* pupae and their downstream ex vivo preparation for live imaging and immunostaining. This strategy enables straightforward access to physiologically relevant innate immune cells, both circulating and tissue-resident populations, for subsequent imaging of signal transduction.

Key words Ex vivo, *Drosophila melanogaster*, Innate immune cell, Macrophage, Immunostaining, Confocal microscopy, Live time-lapse imaging

1 Introduction

In recent decades, studies on immune cell behavior and signaling using the fruit fly *Drosophila* have made a tremendous impact on the field of innate immunity [1, 2]. The optical translucency and relative immobility of *Drosophila* embryos has enabled high resolution in vivo live imaging of fly macrophage-like cells ("hemocytes") within intact living animals [2]. This exceptional capacity for detailed spatiotemporal imaging of immune behavior in vivo, together with *Drosophila*'s renowned genetic tractability, has enabled researchers to dissect the complex signaling networks controlling macrophage migration [3–5]. More recently, methods have been developed for noninvasive time-lapse imaging of macrophages within the later pupal stage of *Drosophila* development, when macrophages are present in far greater numbers and genetic misexpression systems (e.g., GAL4-UAS) are more effective [6–

Christoph Wülfing and Robert F. Murphy (eds.), *Imaging Cell Signaling*, Methods in Molecular Biology, vol. 2800,
https://doi.org/10.1007/978-1-0716-3834-7_1,

10]. Pupae can easily be prepared for live imaging by careful removal of the opaque pupal case with fine forceps to reveal the optically translucent pupa underneath, which can be kept alive with ease for >24 h without obvious detriment [6]. Lately, more sophisticated methods have been developed that enable immobilization of *Drosophila* larvae and adults for long-term live imaging of numerous tissues in vivo (including the gut, nervous system, and macrophages) through carefully produced "windows" in the epidermis [11, 12].

Nevertheless, despite the ease of live imaging within intact *Drosophila,* certain downstream applications must be performed on cells ex vivo. A number of immortalized cell lines exist that are derived from *Drosophila* embryonic macrophages, such as the widely used Schneiders S2 cells [13], and these are used extensively in biomedical research due to the comparative ease of performing RNAi, CRIPSR, and high-throughput screens [14]. Nevertheless, the derived nature of these cell lines limits their usefulness for understanding immune biology in vivo; rather, this necessitates the careful and fresh isolation of primary (living) macrophage cells. Optimized methods already exist for flow cytometry-based isolation of fluorescently labeled fly macrophages using fluorescence-activated cell sorting (FACs) [15], and these have proved invaluable for macrophage expression-based analyses (e.g., via quantitative reverse transcription polymerase chain reaction (RT-qPCR), bulk RNA sequencing, and more recently single-cell RNA sequencing) [16]. These FACs-based methods are particularly valuable when large numbers of macrophages must be collected for sensitive downstream applications that require significant amounts of material (e.g., for RT-qPCR). However, established methods to prepare *Drosophila* for FACs are still comparatively lengthy, and despite methods (e.g., keeping on ice) during sample preparation to minimize cellular damage and maximize cellular survival, it is inevitable that macrophages will suffer damage or altered biochemistry during the procedure; typical macrophage viability rates detected during FACs-based sorting are in the range ~ 30–60% [17]. While it is comparatively straightforward to isolate immature macrophages by simple "bleeding" of larvae and newly formed (<1 h old) pre-pupae [18–20], isolating mature macrophages on bulk for ex vivo analysis from pupae that possess more complex tissue architecture has traditionally been challenging as macrophages are more intimately associated with tissues.

Here, we describe an alternative cost-effective method that enables more rapid and gentle isolation of living macrophages from *Drosophila* pupae of various ages, without the need for lengthy flow cytometry-based detection. We demonstrate that macrophages can be efficiently isolated from entire pupae as well as from specific pupal tissues of interest. Once isolated, ex vivo macrophages can be kept alive and imaged in situ, or processed further for downstream biochemical and biophysical analyses (e.g., immunostaining). Unlike their in vivo counterparts, ex vivo macrophages are more accessible and amenable to cutting-edge (e.g., super-resolution) microscopy methods and biophysical analyses.

2 Materials

Prepare all solutions using ultrapure water (prepared by purifying deionized water) and analytical grade reagents. Prepare and store all reagents at room temperature (unless indicated otherwise).

2.1 Drosophila Husbandry and Pupal Staging

1. Standard fly food media: Typically, a cornmeal-molasses-agar mixture containing maize, soya flour, malt extract, molasses, agar, propionic acid, and nipagen. https://cshprotocols.cshlp.org/content/2012/11/pdb.rec072488.full?rss=1
2. *Drosophila* stocks: All *Drosophila* genotypes described here are readily available from international stock centers, such as the Bloomington Stock Centre, Indiana University https://bdsc.indiana.edu/
3. Apparatus for pupae staging and collection: Fine sable paintbrush, forceps, transparent plastic or glass vials for *Drosophila* culture.

2.2 Ex Vivo Culture of Drosophila Macrophages

1. Supplemented Schneider's Medium (SSM): Schneider's Medium with 10% BSA. Make fresh before each use. Add 10-mL Schneider's Medium (with bicarbonate and L-glutamine) to a sterile 20-mL (flat bottomed) tube. Weigh 1-g BSA and transfer to the tube. Mix gently until BSA has dissolved fully.
2. Imaging dishes with concanavalin A (conA) coating: Glass bottom imaging chamber with concanavalin A coating (conA solution recommended concentration 0.5 mg/ml in 1xPBS). A thin glass suitable for imaging is recommended (e.g., 175 μm).
3. Dissection apparatus: Fine forceps, double-sided tape, glass microscope slides, dissecting stereo microscope with bright-field, micro scissors, small sterile surgical blade, Eppendorf tubes.

2.3 Fixation and Immunostaining

1. Fixation solution: Solution containing 4% formaldehyde in phosphate-buffered saline (PBS). Mix and store at 4 °C for one week. Alternatively, solution can be made ahead of time and stored long term at −20 °C.
2. PBS-TX-BSA blocking solution: PBS solution containing 0.1% Triton X-100 detergent and 1% bovine serum albumin (BSA).

3. Antibodies and fluorescent stains: Staining solutions containing appropriate dyes and antibodies to be made fresh in PBS-TX-BSA just prior to sample application. Protocol can be adapted to label multiple cellular features using dyes such as DAPI (at recommended 1:1000 dilution) for nuclear labeling and fluorescent conjugates of phalloidin (at recommended 1: 200 dilution) for Actin cytoskeletal labelling, as well as fluorescently conjugated primary antibodies against specific cellular epitopes.
4. Media for sample mounting and imaging: PBS can be used for short-term storage, but a solution of Mowiol containing 2.5% 1,4-diazobicyclo-[2.2.2]-octane (DABCO) is recommended for longer-term storage. https://cshprotocols.cshlp.org/content/2006/1/pdb.rec10255

3 Methods

Carry out all procedures at room temperature unless otherwise specified.

3.1 Preparation and Staging of Living Drosophila Pupae

1. Obtain *Drosophila* of the appropriate genotype (*see* **Note 1**).
2. Grow adult flies on standard *Drosophila* food.
3. Collect healthy adult flies of the appropriate genotype (approximately 2–3 days after eclosion).
4. Select healthy adult flies by briefly anesthetizing using carbon dioxide gas pads. A fine paintbrush can be used to transfer flies of the appropriate genotype or gender to a collection vial.
5. Approximately 20 young healthy female and 20 male flies should be added to each vial containing standard fly food media supplemented with fresh yeast (*see* **Note 2**).
6. Select newly formed white pre-pupae (at 0 h APF, after puparium formation). This should be performed at a predetermined time, in advance of the planned dissection procedure. For example, if macrophages are to be isolated from pupae that are 18 h APF, pre-pupae collection should be performed 18 h before the scheduled dissection session (*see* **Note 3**).
7. The selected white pre-pupae should be carefully removed from the vials and transferred to the side of a fresh vial (*see* **Notes 4** and **5**).
8. The selected pre-pupae should now be aged to the appropriate developmental stage (e.g., 18 h APF or 75 h APF) in the vial at 25 °C.

3.2 Preparation of Imaging Chamber for Seeding Primary Drosophila *Cells*

1. Prior to pupal dissection, a glass bottomed imaging chamber should be coated with concanavalin A. For this, we recommend adding 0.5 mL of 0.5 mg/mL concanavalin A solution to the glass coverslip surface for 30 min (*see* **Note 6**).
2. Excess conA solution should be removed from the imaging chamber, air-dried, and the glass surface briefly washed with PBS.
3. An appropriate volume of Supplemented Schneider's Medium (SSM, we recommend 1.5 mL) should be added to the coated imaging dish and the dish maintained at room temperature (*see* **Note 7**).
4. SSM should also be added to an Eppendorf tube or equivalent container where both the tissue and your forceps can be immersed comfortably.

3.3 Pupal Dissection and Macrophage Isolation

1. Carefully transfer the staged *Drosophila* pupae to a piece of double-sided sticky tape attached to a glass slide. Each pupa should be positioned with the ventral side stuck to the tape and the dorsal side facing upward. The anterior of each pupa will be easily identifiable from the two spiracles protruding from the anterior-most end of the pupal case.
2. Using a brightfield (or fluorescent, if available) dissection microscope, carefully remove each pupa from its protective case (puparium) using forceps. Micro scissors can also be used to aid the dissection (*see* **Note 8**).
3. If macrophages residing in specific pupal tissues are to be isolated, use forceps to quickly remove the specific pupal tissue of interest (e.g., the pupal wings). Tissues should be briefly washed in SSM (pre-prepared in the Eppendorf) to remove any liquid (e.g., hemolymph) associated with the tissue, and the tissue then transferred to the SSM within the prepared imaging chamber (*see* **Note 9**). Alternatively, whole pupae can be processed for isolation of both whole-body circulating macrophage populations and tissue-resident macrophages.
4. After transferring the pupae or tissue of interest, it is important to ensure the tissue is fully submerged in the media.
5. Once all pupal tissues of interest have been transferred to the imaging chamber, tissues should be mechanically dissociated in situ, while submerged in SSM, using sterile forceps. The aim of this mechanical dissociation is to release macrophages from their associated (e.g., epithelial) tissues (Fig. 1).
6. Macrophages should then be allowed to adhere to the conA-treated glass surface for 20 min at room temperature (*see* **Note 10**).

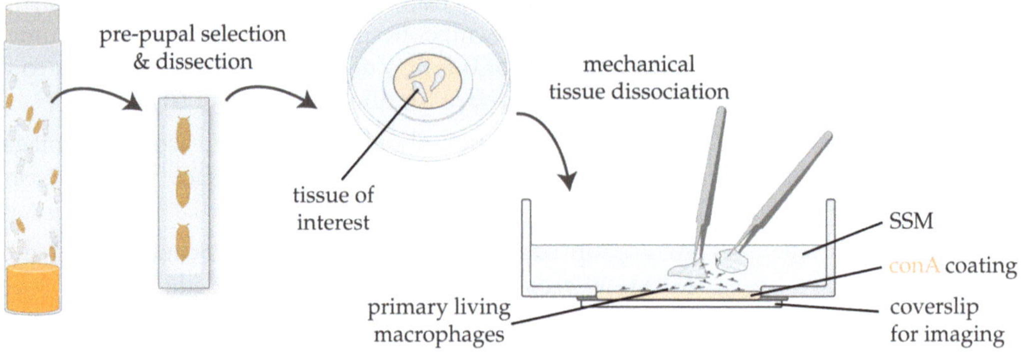

Fig. 1 Schematic showing setup for *Drosophila* pupal tissue dissociation and isolation of living primary macrophages for downstream ex vivo applications

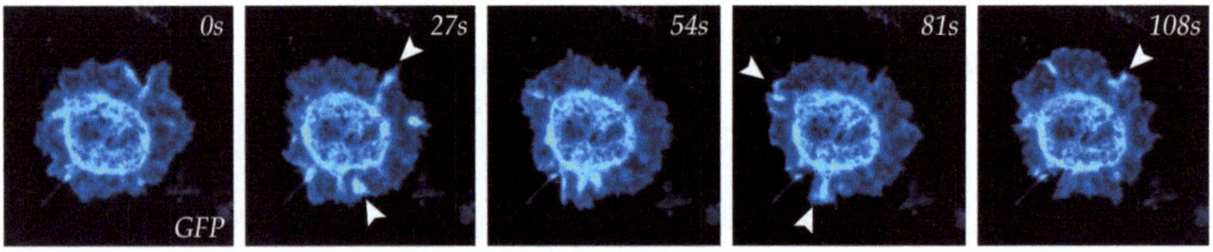

Fig. 2 Time-lapse confocal imaging of living primary pupal macrophages expressing cytoplasmic GFP using the macrophage-specific serpent promoter. Macrophages exhibit dynamic lamellipodial protrusions

7. Macrophages can be live-imaged at this stage if further processing is not required (Fig. 2). At this point in the protocol, macrophages should be well adhered to the coverslip, and unwanted tissue debris can be gently rinsed off using SSM.

3.4 Processing of Primary Macrophages for Downstream Ex Vivo Applications

1. Prior to staining, macrophages must be chemically fixed. For macrophage fixation, the SSM media should be removed from the imaging chamber, taking care not to disturb the adhered macrophages (*see* **Note 11**).
2. Fixative solution (4% formaldehyde in PBS) should be carefully pipetted onto the macrophages and incubated at room temperature for 15 min without agitation (*see* **Note 12**).
3. The fixation solution should then be carefully removed from the chamber and the cells gently washed three times for 10 min with PBS solution on an orbital shaker.
4. The PBS solution should be removed and replaced with PBS-TX-BSA solution that acts both to permeabilize the cells and as a blocking buffer to reduce nonspecific binding. Macrophages should be incubated with PBS-TX-BSA solution for 30 min on an orbital shaker.
5. PBS-TX-BSA solution should then be replaced with a solution of PBS-TX-BSA containing the appropriate concentration of

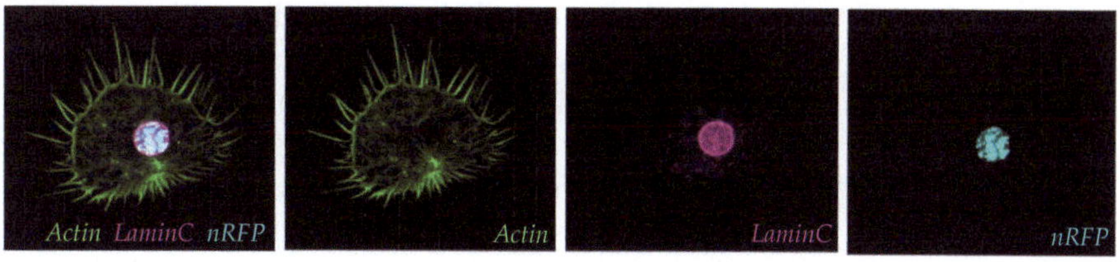

Fig. 3 Confocal microscopy images of primary pupal macrophages expressing nuclear RFP (nRFP, cyan) following fixation and staining to label Actin (phalloidin, green) and the nuclear lamina (anti-LaminC, magenta)

fluorescent conjugate (e.g., DAPI or phalloidin) and incubated for 1 h at room temperature on an orbital shaker (*see* **Note 13**).

6. If antibody staining is to be performed, we recommend incubating the adhered cells first with PBS-TX-BSA solution containing the primary antibody for 2 h at room temperature while being agitated (e.g., using an orbital shaker). Primary antibody incubation should then be followed by three washes of 15 min each in PBS solution on an orbital shaker before addition of the secondary antibody solution and incubation overnight at 4 °C. Antibody solution should then be removed before performing three washes of 15 min each in PBS. At this point, further staining (e.g., with DAPI or phalloidin) can be performed as recommended above.
7. Following staining, the staining solution should be removed and adhered cells carefully washed three times with PBS for 15 min on the orbital shaker.
8. Stained cells can be left in PBS in the short term. For long-term storage, PBS should be removed and replaced with Mowiol solution (to which anti-fade has been added) and stored at 4 °C until imaging. At this stage, wrapping film can be used around the lid to prevent dehydration.
9. Stained macrophages can then be imaged using appropriate fluorescent microscopes, for example, confocal microscopy (Fig. 3).
10. Microscopy images can be visualized either during image capture or later with specialist image analysis software (such as ImageJ, Volocity or IMARIS) using z-stack projections or 3D rendering.

4 Notes

1. This protocol specifically utilizes the pupal stage of *Drosophila* development. Consequently, this protocol is only suitable for *Drosophila* genotypes that are viable until at least late pupal

stages; embryonic lethal stocks will not be suitable. This is unlikely to be a problem when imaging control (wild-type) pupae but must be an important consideration when isolating macrophages from transgenic pupae (in which genes are knocked down or overexpressed perhaps via the GAL4-UAS genetic misexpression system [21]). In the case of early lethality caused by GAL4-mediated genetic manipulation, a Gal80^{ts} construct can be used in combination with the GAL4-UAS system to inhibit unwanted GAL4 activity early in development.

2. If the Gal4-UAS system is being used to drive macrophage-specific gene expression, *Drosophila* must be raised at 25 °C or above because the Gal4-UAS system is highly temperature sensitive and will not be effective at temperatures below 25 °C.
3. For optimal pupal collection, the adult (parental) flies should be tipped each day onto new food in fresh vials, keeping all vials at 25 °C. Flies can also be tipped into larger food bottles if considerable numbers of pupae are needed, for example, if a specific genotype needs to be selected. It is advisable to time the prepupae collection in line with the anticipated start of macrophage extraction from the tissue; for example, if pupal case removal is expected to take a long time (e.g., more than 30 min), account for this additional time by adding 30 min to the prepupae selection time.
4. Mature (wandering third instar) larvae crawl upward out of the media to undergo pupation; newly formed white prepupae can be identified as they possess everted anterior spiracles and are stationary (unlike third instar larvae). At the start of pupation, the larval cuticle transforms into a "puparium" (the pupal case). The pupal case is initially soft and white but becomes progressively darker and more brittle as the pupa develop.
5. Forceps or a fine paintbrush are suitable for selection and transfer of pupae. During pre-pupae transfer, care should be taken to avoid damaging the pupae. Damage can lead not only to an unwanted injury-induced inflammatory response but also to developmental delay.
6. Addition of conA to the glass chamber increases the adherence of macrophages to the glass imaging surface; this adherence is important for macrophage spreading and to ensure macrophages are retained in the imaging chamber during downstream staining steps.
7. Enough SSM should be added to the dissection/imaging chamber to completely cover your sample, to ensure that the sample isn't disrupted by the liquid surface tension.
8. We have previously published a detailed protocol that describes the steps required to remove the translucent pupa from its

opaque pupal case [6]. Using fine forceps, an incision should be made in the anterior-most region of the puparium; the pupal case in this area is devoid of pupal tissue and is therefore a useful starting point for the dissection. After this initial incision, the pupal case should be peeled gradually open using forceps or micro scissors, working in an anterior to posterior direction until the pupa is completely free of the brittle casing. Forceps can be sharpened to a preferred shape to aid dissection, for example, only sharpening the side facing the pupal case when placed between the case and the pupae. Once removed from their cases, pupae are very fragile, and care must be taken to avoid puncturing the pupal surface. Puncturing will result in fly blood (hemolymph) quickly leaking out from the puncture site. Pupae with punctures, however small, should be discarded.

9. Rinsing the tissue in SSM will remove any contaminant cells or tissues (e.g., circulating macrophages in the hemolymph) that may be associated with the tissue of interest. Maintain a firm grip with the forceps on your tissue of interest while rinsing in SSM; for example, for pupal wings, a tight grip prevents hemolymph from within the wings draining out into the SSM. A small surgical blade may be useful in excising the tissue while it is pinched with forceps.
10. Unlike larval macrophages, pupal macrophages will polarize and will quickly spread (evident by increased cell size) on the conA-coated glass surface.
11. Use a pipette tip with a wide tip, such as P1000. Align the pipette tip at the side of the imaging dish and remove the liquid extremely slowly. It is preferrable to leave a small amount of SSM at the bottom of the imaging dish, as breaking the water surface tension will create disturbing water flows.
12. Add the fixative to the middle of the dish one drop at the time, making sure that it spreads evenly as it mixes with any remaining SSM. Do not move the imaging dish during the fixation.
13. For DAPI stains, a shorter 15-min incubation will be sufficient. At this stage, antibody concentrations and incubation times found in standard cell culture protocols will generally be appropriate.

Acknowledgments

This work was supported by Wellcome Trust and Royal Society Sir Henry Dale Fellowship 208762/Z/17/Z to H.W.

References

1. Weavers H, Martin P (2020) The cell biology of inflammation: from common traits to remarkable immunological adaptations. J Cell Biol 219. https://doi.org/10.1083/JCB.202004003
2. Weavers H, Wood W (2016) Creating a buzz about macrophages: the fly as an in vivo model for studying immune cell behavior. Dev Cell 38:129–132. https://doi.org/10.1016/j.devcel.2016.07.002
3. Akhmanova M, Emtenani S, Krueger D et al (2022) Cell division in tissues enables macrophage infiltration. Science 376:394. https://doi.org/10.1126/SCIENCE.ABJ0425
4. Yolland L, Burki M, Marcotti S et al (2019) Persistent and polarized global actin flow is essential for directionality during cell migration. Nat Cell Biol 21:1370–1381. https://doi.org/10.1038/s41556-019-0411-5
5. Weavers H, Evans IR, Martin P, Wood W (2016) Corpse engulfment generates a molecular memory that primes the macrophage inflammatory response. Cell 165:1658–1671. https://doi.org/10.1016/J.CELL.2016.04.049
6. Weavers H, Franz A, Wood W, Martin P (2018) Long-term in vivo tracking of inflammatory cell dynamics within *Drosophila* pupae. J Vis Exp. https://doi.org/10.3791/57871
7. Liepe J, Sim A, Weavers H et al (2016) Accurate reconstruction of cell and particle tracks from 3D live imaging data. Cell Syst 3:102–107. https://doi.org/10.1016/j.cels.2016.06.002
8. Weavers H, Liepe J, Sim A et al (2016) Systems analysis of the dynamic inflammatory response to tissue damage reveals spatiotemporal properties of the wound attractant gradient. Curr Biol 26:1975–1989. https://doi.org/10.1016/j.cub.2016.06.012
9. Thuma L, Carter D, Weavers H, Martin P (2018) *Drosophila* immune cells extravasate from vessels to wounds using Tre1 GPCR and Rho signaling. J Cell Biol 217:3045–3056. https://doi.org/10.1083/jcb.201801013
10. Sander M, Squarr AJ, Risse B et al (2013) *Drosophila* pupal macrophages – a versatile tool for combined ex vivo and in vivo imaging of actin dynamics at high resolution. Eur J Cell Biol 92:349–354. https://doi.org/10.1016/J.EJCB.2013.09.003
11. Martin JL, Sanders EN, Moreno-Roman P et al (2018) Long-term live imaging of the *Drosophila* adult midgut reveals real-time dynamics of division, differentiation and loss. elife 7. https://doi.org/10.7554/ELIFE.36248
12. Kakanj P, Eming SA, Partridge L, Leptin M (2020) Long-term in vivo imaging of *Drosophila* larvae. Nat Protoc 15:1158–1187. https://doi.org/10.1038/S41596-019-0282-Z
13. Schneider I (1972) Cell lines derived from late embryonic stages of *Drosophila* melanogaster. Development 27:353–365. https://doi.org/10.1242/DEV.27.2.353
14. Buster DW, Nye J, Klebba JE, Rogers GC (2010) Preparation of *Drosophila* S2 cells for light microscopy. J Vis Exp 40. https://doi.org/10.3791/1982
15. Gyoergy A, Roblek M, Ratheesh A et al (2018) Tools allowing independent visualization and genetic manipulation of *Drosophila melanogaster* macrophages and surrounding tissues. G3: Genes Genomes Genet 8:845–857. https://doi.org/10.1534/G3.117.300452/-/DC1
16. Ramond E, Dudzic JP, Lemaitre B (2020) Comparative RNA-Seq analyses of *Drosophila* plasmatocytes reveal gene specific signatures in response to clean injury and septic injury. PLoS One 15:e0235294. https://doi.org/10.1371/JOURNAL.PONE.0235294
17. Clemente GD, Weavers H (2023) A PI3K-calcium-Nox axis primes leukocyte Nrf2 to boost immune resilience and limit collateral damage. J Cell Biol 222. https://doi.org/10.1083/JCB.202203062
18. Sampson CJ, Williams MJ (2012) Real-time analysis of *Drosophila* post-embryonic Haemocyte behaviour. PLoS One 7:e28783. https://doi.org/10.1371/JOURNAL.PONE.0028783
19. Rüder M, Nagel BM, Bogdan S (2018) Analysis of cell shape and cell migration of *Drosophila* macrophages in vivo. Methods Mol Biol 1749: 227–238. https://doi.org/10.1007/978-1-4939-7701-7_17
20. Parton RM, Vallés AM, Dobbie IM, Davis I (2010) *Drosophila* macrophage preparation and screening. Cold Spring Harb Protoc 2010:pdb.prot5404. https://doi.org/10.1101/PDB.PROT5404
21. Brand AH, Perrimon N (1993) Targeted gene expression as a means of altering cell fates and generating dominant phenotypes. Development 118:401–415. https://doi.org/10.1101/lm.1331809

Chapter 2

Generation of 3D Fibroblast-Derived Extracellular Matrix and Analysis of Tumor Cell-Matrix Interactions and Signaling

Emily Lay, Tressan Grant, Gernot Walko, and Ute Jungwirth

Abstract

Fibroblasts are the major producers of the extracellular matrix and regulate its organization. Aberrant signaling in diseases such as fibrosis and cancer can impact the deposition of the matrix proteins, which can in turn act as an adhesion scaffold and signaling reservoir promoting disease progression. To study the composition and organization of the extracellular matrix as well as its interactions with (tumor) cells, this protocol describes the generation and analysis of 3D fibroblast-derived matrices and the investigation of (tumor) cells seeded onto the 3D scaffolds by immunofluorescent imaging and cell adhesion, colony formation, migration, and invasion/transmigration assays.

Key words Extracellular matrix, Matrix deposition assay, Fibroblasts, Cell-derived matrix, Fibronectin, Cell-matrix interaction, Matrix adhesion

1 Introduction

The extracellular matrix, which can also be referred to as the matrisome, is a sophisticated three-dimensional arrangement of collagens, proteoglycans, and glycoproteins (core components) as well as associated proteins, regulators, and secreted factors. It provides a physical scaffold for cellular adhesion and migration, as well as mechanical and biochemical cues that mediate cell signaling and consequently control cell proliferation and survival, as well as cell differentiation and lineage specification. The extracellular matrix is a highly dynamic structure, and its composition and organization change during aging, wound healing, and tumor progression, thereby impacting disease pathogenesis and therapy responses [1, 2]. Fibroblasts are the major cell type that deposits and remodels the extracellular matrix. Activation of fibroblasts is a key feature in fibrosis and cancer and is associated with extensive alterations of

Christoph Wülfing and Robert F. Murphy (eds.), *Imaging Cell Signaling*, Methods in Molecular Biology, vol. 2800,
https://doi.org/10.1007/978-1-0716-3834-7_2,

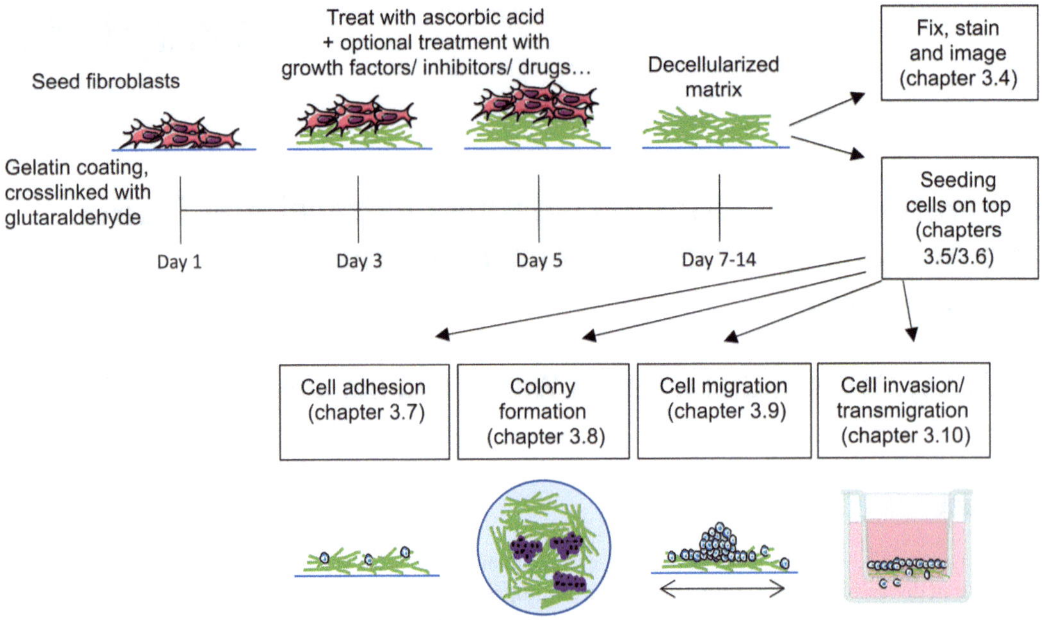

Fig. 1 Graphical outline of the described methodology

the extracellular matrix, often leading to stiffening of the (tumor) microenvironment [3]. With the identification of different cancer-associated fibroblast subpopulations, it has become clear that unique matrisome signatures are expressed by the distinct fibroblast subtypes impacting the cross talk with cancer cells and their cellular signaling [4, 5].

In this protocol, we outline a method of generating and assessing 3D fibroblast-derived extracellular matrices in cell culture dishes. These 3D cell-derived matrices enable the investigation of matrix composition [6] and orientation upon different stimuli, as well as (tumor) cell-matrix interactions and signaling (Fig. 1).

The method is based on providing ascorbic acid to fibroblasts, which stimulates cell proliferation and deposition of extracellular matrix proteins, particularly fibronectin and collagens [7]. A glutaraldehyde-cross-linked gelatin layer creates a bioadhesive surface for anchoring the fibroblast-derived matrix [8]. To investigate the cross talk between fibroblasts and tumor cells, fibroblasts can be treated with recombinant signaling molecules (e.g., TGFβ), non-cytotoxic drugs, or conditioned medium from tumor cells while depositing the matrix. After the fibroblasts have deposited their own extracellular matrix, it can then be decellularized and investigated by immunofluorescent antibody staining and microscopy. Fiber alignment can be examined by image analysis (Subheadings 3.1, 3.2, 3.3 and 3.4) (Fig. 2). Besides being able to assess how different signaling impacts the matrix itself, (tumor) cells and

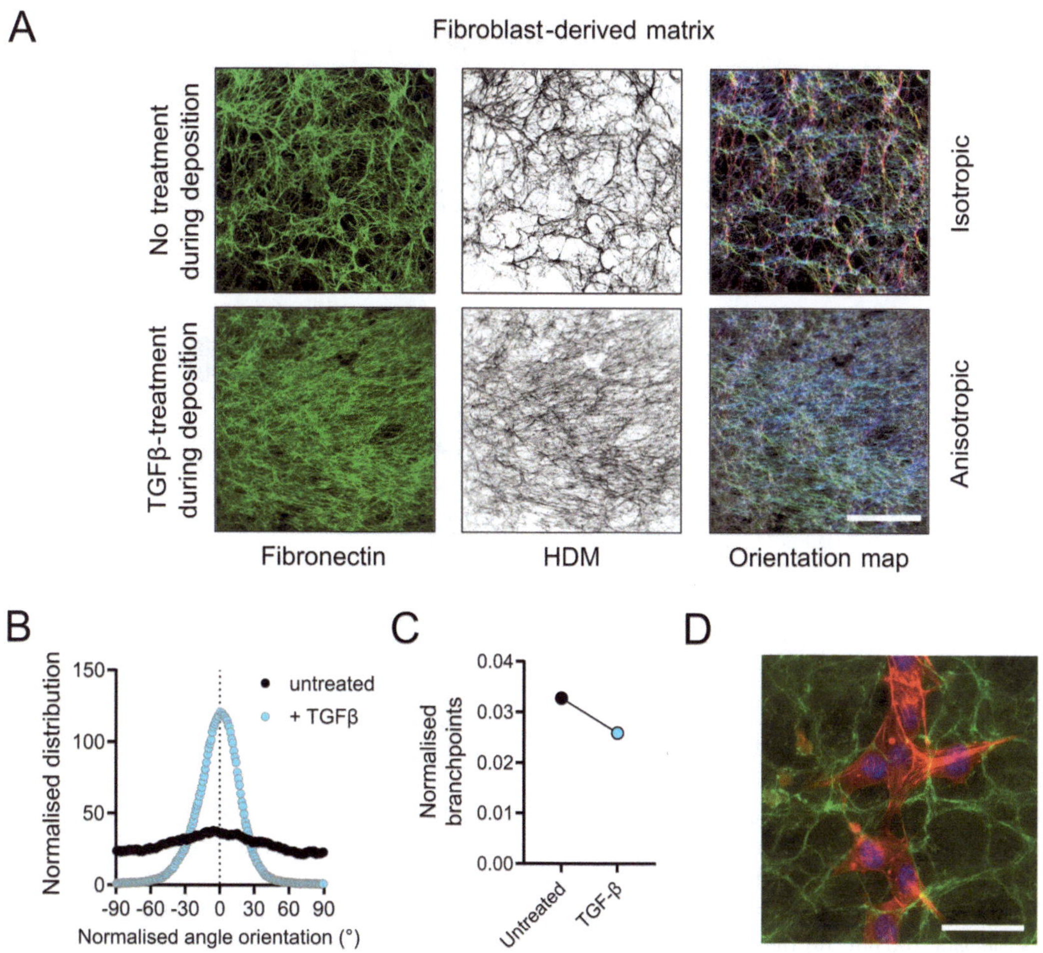

Fig. 2 Visualization and analysis of fibronectin in fibroblast-derived matrix. (**a**) Extracellular matrix derived from untreated or TGFβ-treated fibroblasts, immunostained for fibronectin. Left panel, immunofluorescence confocal image. Middle panel, high-density matrix projection (HDM). Right panel, colored fiber orientation map (scale bar, 50 μm). Example quantification of (**b**) angle orientation using OrientationJ [9] and of (**c**) normalized branch points using TWOMBLI [10]. (**d**) Tumor cells (DAPI, blue; phalloidin, red) growing on a fibroblast-deposited matrix (fibronectin, green) (scale bar, 50 μm)

fibroblasts can be reseeded on top of the matrices (Subheadings 3.5 and 3.6), and their behavior can be monitored either live or upon fixation and immunofluorescent antibody staining for signaling proteins. Furthermore, we also provide protocols for imaging (tumor) cell adhesion (Subheadings 3.7), colony formation (Subheadings 3.8), and (tumor) cell migration (Subheadings 3.9) on top of the fibroblast-derived matrices (Fig. 3), as well as tumor cell invasion/transmigration (Subheadings 3.10) through the cell matrix.

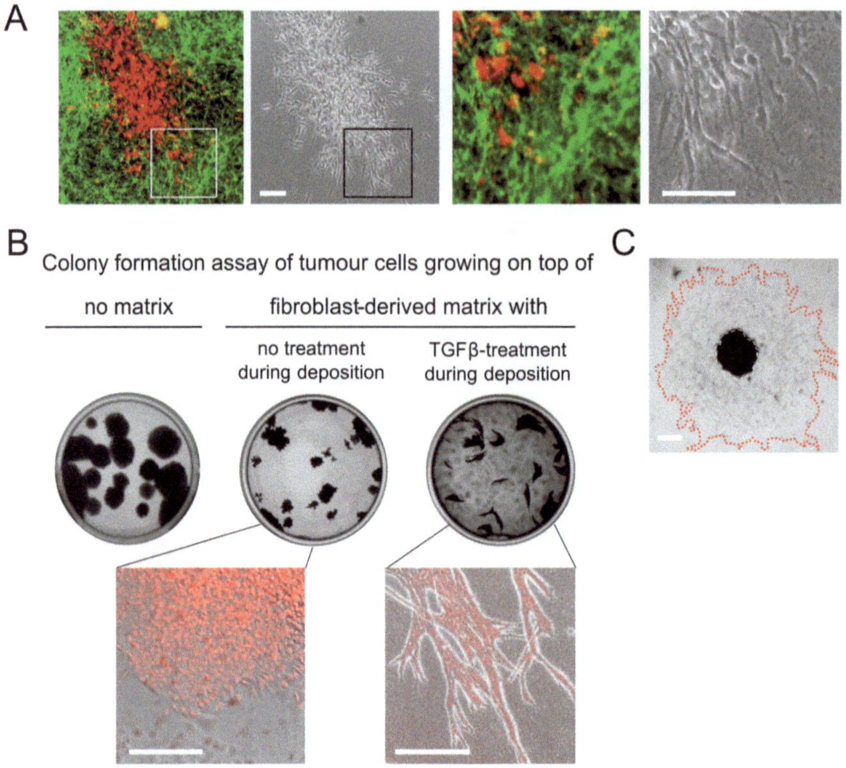

Fig. 3 Colony formation and cell migration on top of fibroblast-derived matrix. (**a**) Colony formation assay. Left panels: RFP-labelled tumor cells were cultured for 7 days on top of a fibroblast-derived matrix (fibronectin, green) in a 24-well format, and imaged with fluorescence and brightfield/phase contrast optics. Right panels: magnified areas of left panels. Note that the matrix is visible in the brightfield image at higher magnification (scale bars, 100 μm). (**b**) The impact of fibroblast-derived matrices and signaling molecules such as TGF-β on tumor cell growth. Fibroblasts were treated with TGFβ during the deposition period and tumor cells seeded on top after decellularization. Top panels, crystal violet stained tumor cell colonies. Bottom panels, higher magnification images (brightfield and fluorescence) of tumor cells expressing RFP-tagged histone (scale bars, 250 μm). (**c**) Migration assay. Brightfield image shows outward migration of tumor cells from spheroid placed on top of fibroblast-deposited extracellular matrix. Red line shows front of migrated cells, white line indicates spheroid body (scale bar, 250 μm)

2 Materials

2.1 Cell Culture

1. 24-well cell culture plates and 13-mm glass coverslips for culturing cells (*see* **Note 1**).
2. Fibroblast cell models such as MRC5 (human lung fibroblasts), 3T3 (murine embryonic fibroblasts), or tumor-derived cancer-associated fibroblasts [5], which can form confluent layers. Required cell quantities vary depending on the vessel size used and fibroblast line. For one 24-well plate at least 6×10^5 cells are required.

3. Cell culture medium: Dulbecco's Modified Eagle's Medium (DMEM)-high glucose, supplemented with 10% (v/v) fetal bovine serum (FBS), 2-mM L-glutamine, and 1% penicillin/ streptomycin (*see* **Note 2**).

2.2 Cell Culture Dish/ Coverslip Preparation

1. Pre-treated coverslips: Put coverslips in 1-M HCl and incubate at 60 °C for 4–16 h, with agitation (*see* **Note 3**). Rinse coverslips thoroughly with 100% (v/v) ethanol, followed by overnight incubation in 100% (v/v) ethanol. Store in 100% (v/v) ethanol.
2. Gelatin coating solution: Prewarm viscous gelatin stock in a water bath at 37 °C to liquify. Prepare a 0.2% (v/v) sterile gelatin solution in PBS. Filter-sterilize through a 0.2-μm filter unit and store at 4 °C.
3. Glutaraldehyde crosslinking solution: Prepare a 1% (v/v) sterile glutaraldehyde solution in PBS. Filter-sterilize through a 0.2-μm filter unit. Store at 4 °C.
4. Glycine quenching solution: Prepare a 1-M sterile glycine solution in PBS. Filter-sterilize through a 0.2-μm filter unit. Store at 4 °C.

2.3 ECM Deposition

1. Ascorbic acid solution: Reconstitute ascorbic acid in sterile dH_2O at a concentration of 10 mg/mL. Aliquot and store at −20 °C.
2. 1x concentrated matrix deposition medium: Supplement cell culture medium (2.1.3) with 50 μg/mL ascorbic acid. Prepare fresh.
3. 2x concentrated matrix deposition medium: Supplement cell culture medium (2.1.3) with 100 μg/mL ascorbic acid. Prepare fresh.
4. *Optional*: Recombinant human TGF-β: Reconstitute according to manufacturer instructions. Aliquot and store at −20 °C to −70 °C.

2.4 ECM Decellularization

1. ECM decellularization buffer: 20 -M ammonium hydroxide solution (NH_4OH) supplemented with 0.5% (v/v) Triton X-100 in sterile PBS. For 100 mL add 2 mL of a 1-M NH_4OH to 90 mL PBS, add 0.5-mL Triton X-100, mix. Once dissolved, make up to 100 mL with sterile PBS and store at 4 °C.
2. Washing buffer: Dilute 10x concentrated PBS with Ca^{2+} and Mg^{2+} ("PBS++") with sterile dH_2O to a 1x concentrated PBS++ buffer.
3. DNAse I solution: Prepare a 10 U/mL DNase solution in sterile PBS++.

2.5 Reagents for Fixing and Staining Matrix Proteins and Cells Grown on Top of Matrices

1. *Optional when fixing the matrix:* Fixation solution: 4% (v/v) paraformaldehyde in PBS.
2. *Optional when coverslips are mounted for imaging:* Glass slides and mounting reagent (*see* **Note 4**).
3. *Optional when cells are grown on top of matrices:* Permeabilization buffer: 0.5% (v/v) Triton X-100 in PBS.
5. *Optional for immunostaining:* Blocking and staining buffer: 1% (w/v) BSA and 2% (v/v) FBS in PBS++. Filter-sterilize through a 0.2-μm filter unit and store at −20 °C.
4. *Optional for immunostaining:* Rabbit antihuman fibronectin antibody (or against protein of interest), rabbit-IgG-Alexa Fluor 488 antibody, DAPI, Alexa Fluor546-phalloidin diluted in blocking buffer (typically 1:500–1:2000).
5. *Optional for colony formation assay:* Crystal violet staining solution: 0.4% (w/v) crystal violet dissolved in 25% (v/v) methanol in dH_2O. Store at room temperature.

3 Methods

All solutions and equipment used with living cells must be sterile, and aseptic techniques have to be used. Some of the chemicals in this protocol, for example, glutaraldehyde, crystal violet, and paraformaldehyde are hazardous to human health.

3.1 Preparation of Coverslips/Cell Culture Dishes for Matrix Deposition Essay

1. *Optional:* Using sterile forceps, dip sterilized coverslips stored in ethanol (*see* Subheading 2.1) into PBS and place into wells of a 24-well cell culture plate.
2. Coat cell culture wells (or coverslips or transwells) with gelatin by adding 250 μL of sterile 0.2% (v/v) gelatin coating solution into each well and incubate for 60 min at 37 °C (*see* **Note 5**).
3. Aspirate the gelatin solution and wash each well three times with at least 250 μL PBS.
4. To cross-link gelatin, add 250 μL of the sterile 1% (v/v) glutaraldehyde cross-linking solution to each well and incubate for 30 min at room temperature (*see* **Note 6**).
5. Remove the glutaraldehyde solution (*see* **Note** 7).
6. Wash each well three times with at least 250 μL PBS for 5 min.
7. To quench the cross-linking, add 250 μL of the sterile 1-M glycine quenching solution in PBS to each well and incubate for 20 min at room temperature.
8. Aspirate the glycine solution and wash each well three times with at least 250 μL PBS. Use immediately according to 3.2 or store at 4 °C (*see* **Note 8**).

3.2 Matrix Deposition

1. Before using the (coated) cell culture plates or coverslips, incubate them with cell culture medium for 30 min in an incubator (37 °C, 5% CO_2).
2. Detach fibroblasts with trypsin/EDTA solution following the appropriate protocol for the respective cell line/primary cells. Collect them in cell culture medium supplemented with serum. Count cells and dilute appropriately. For seeding into a 24-well plate, a concentration of 2.5–5 × 10^4 cells/mL is recommended. Add 1 mL of the cell suspension per 24-well or coverslip (*see* **Note 9**).
3. Culture fibroblast overnight in an incubator (37 °C, 5% CO_2). Before the experiment is continued, the cells should have formed a confluent layer, which should be maintained throughout the matrix deposition. If cells are not confluent, grow for an additional day.
4. When the cells have formed a confluent layer, gently aspirate the growth medium from each well. Replace with 1 mL 1x concentrated matrix deposition medium. *Optional:* In addition to the ascorbic acid contained in the matrix deposition medium, fibroblast can be treated with growth factors such as TGF-β or other factors and drugs.
5. As ascorbic acid is unstable in cell culture medium, it has to be replaced every second day for at least 7–14 days. To avoid disturbing the fibroblast cell layer and matrix generation, it is recommended to remove only half (0.5 mL) of the culture medium and to replace it with 0.5 mL of 2x concentrated matrix deposition medium (*see* **Note 10**).

3.3 Matrix Decellularization

1. Warm the decellularization buffer to 37 °C.
2. Carefully aspirate, using a pipette, 900 μL of the culture medium, and wash two times with 500 μL PBS (*see* **Note 11**). After the final wash, remove as much of the PBS as possible without disturbing the matrix.
3. Add 250 μL of the pre-warmed decellularization buffer to each well and incubate at 37 °C. Cell lysis will happen rapidly and in most cases be complete within 2–5 min. Use an inverted microscope to confirm that fibroblasts are lysed. If there are remaining fibroblasts, continue cell lysis for 1 min and then check again under the microscope (*see* **Note 11**).
4. To wash the cellular debris from the matrices, gently add 250 μL of PBS++ without removing the extraction buffer from the first step (Subheading 3.1). Subsequently, remove 250 μL of buffer and add again fresh 250 μL PBS++ in each well, avoiding any turbulences. Repeat this step two more times to wash the well.

5. Gently remove as much PBS++ as possible (*see* **Note 11**) and add 250 μL DNase I buffer to digest residual DNA that will have leaked out of the lysed cells. Incubate for 30 min at 37 °C in a cell culture incubator (5% CO_2).
6. Gently remove DNase I buffer and wash twice with PBS++. You should be able to see the matrix under a microscope, using phase contrast optics with a 10x or 20x objective.
7. Either directly continue with matrix fixation and staining (Subheading 3.4), cellular assays (Subheading 3.5, 3.6, 3.7, 3.8, 3.9 and 3.10) on top of matrices or store matrices at 4 °C for up to 2 months. For this, add at least 1 mL PBS++ to each well supplemented with 1% penicillin/streptomycin.

3.4 Optional: Matrix Fixation, Staining, and Image Analysis

1. To fix the matrices, carefully remove the PBS++, add 4% (v/v) paraformaldehyde fixation solution to each well, incubate at room temperature for 20 min and collect paraformaldehyde (*see* **Note** 7). Carefully wash two times with 250 μL PBS++.
2. Aspirate all PBS++ with a pipette and add 250 μL blocking buffer. Incubate for 1 h at room temperature.
3. For fibronectin staining (or alternative protein of interest), remove the blocking buffer and add 250 μL of the primary antibody diluted in blocking buffer to each well). Incubate at room temperature for 1 h (*see* **Notes 12** and **13**).
4. Gently remove the primary antibody solution from each well and carefully wash three times with 250 μL PBS for 5 min at room temperature.
5. For the incubation with the secondary antibody, add 250 μL of the appropriate secondary antibody diluted in blocking buffer. Incubate in the dark at room temperature for 40 min (*see* **Note 14**).
6. Gently remove the secondary antibody solution from each well and carefully wash three times with 250 μL PBS for 5 min (*see* **Note 15**).
7. If matrices were generated on glass bottom plates, these can be directly imaged on an inverted fluorescent microscope. If cells were cultured on coverslips, follow the next steps to mount coverslips on glass slides for imaging.
8. Add a droplet of 5 μL mounting medium (*see* **Note 4**) onto a glass slide. Avoid creating bubbles.
9. To remove the coverslips from the wells, use a needle with a bent tip to grasp the edges of the coverslips and lift them off while grabbing them gently with forceps.

10. Dip the coverslips in dH_2O and then gently tap the edges of the coverslips on a piece of filter paper to remove any excess liquid.
11. Gently and slowly place the coverslips with the matrix side facing downward onto the mounting medium, carefully avoiding the trapping of any air bubbles.
12. *Optional:* If non-curing mounting medium is used, seal around the coverslips using clear nail varnish. Once dry, the slides can be stored in the dark at 4 °C for short-term and at −20 °C for long-term storage.
13. Images can be taken on any fluorescent microscope, such as a confocal laser scanning microscope, high-content imaging system, or a standard fluorescent microscope. Sample images show fibronectin-stained matrices imaged on a confocal microscope with a 40x objective (Fig. 2a).
14. For fibronectin fiber analysis, images are converted to 8-bit images and loaded into FIJI (ImageJ, https://imagej.net/). Fiber alignment can be analyzed using either the OrientationJ plugin (http://bigwww.epfl.ch/demo/orientation/) [9] or the TWOMBLI plugin (https://github.com/wershofe/TWOMBLI) [10]. OrientationJ provides a visual representation of an orientation mask and a vector field map and quantitative data on fiber angles and distribution (Fig. 2a, b). TWOMBLI provides a visual representation of matrix masks and quantification of fiber endpoint number, curvature, fractal dimension, branchpoint number, and matrix density (Fig. 2a, c). Analysis was conducted according to the provided instructions in the supporting documentation of the publications.

3.5 Optional: Preparing Matrices for Growing Cells on Top

1. Gently wash the unfixed matrices twice with PBS++ (*see* **Note 16**).
2. *Optional*: It is recommended to pre-incubate the decellularized matrices for at least 1 h with the specific cell culture medium that will be used in subsequent assays.
3. Detach (tumor) cells with trypsin/EDTA solution following the appropriate protocol for the respective cell line/primary cells and collect them in cell culture medium supplemented with serum. Count cells and dilute appropriately (*see* **Note 17**).
4. For culturing cells on top of matrices, continue with Subheading 3.6, for matrix-tumor cell adhesion assay, continue with Subheading 3.7, for colony formation assay, continue with Subheading 3.8, for tumor cell migration on top of matrices, continue with Subheading 3.9, and for tumor cell invasion through matrices, continue with Subheading 3.10.

3.6 Optional: Culturing (Tumor) Cells on Top of Matrices for Imaging

1. In a 24-well format, gently seed 2×10^4–5×10^4 cells/well in 1 mL cell culture medium on top of the fibroblast-derived matrix (*see* **Note 11**). Incubate for 24–72 h at 37 °C in a cell culture incubator (5% CO_2). Cells can be treated during this period with growth factors, drugs, or medium conditioned from other cell types.
2. Cell proliferation can be monitored by phase contrast imaging (*see* **Note 18**).
3. To fix cells on top of matrices, wash with PBS++ (*see* **Note 19**) and add 4% (v/v) paraformaldehyde fixation solution for 30 min at room temperature. Afterward, collect paraformaldehyde (*see* **Note 7**).
4. Wash wells once with PBS and then permeabilize the cells with 0.5% (v/v) Triton X-100 permeabilization buffer for 10 min at room temperature and wash two times with PBS.
5. For the staining, follow the procedure outlined in 3.4.2–3.4.3. Sample images are shown in Fig. 2d. Alternatively, deposited matrix proteins, cellular proteins, or RNA of the adherent tumor cells can be isolated using standard protocols.

3.7 Optional: Matrix-Tumor Cell Adhesion Assay

1. After preparing the matrices and cells (*see* **Note 20**) according to **step 3.5**, dilute cells appropriately.
2. In a 24-well format, gently pipette 1×10^5 cells in 1 mL growth medium on top of the matrix (*see* **Note 11**). Incubate for 1 h at 37 °C in a cell culture incubator (5% CO_2) (*see* **Note 21**).
3. Non-adherent cells are removed by gently aspirating the cell solution and washing the wells with 1 mL PBS++.
4. The remaining adhered cells are fixed with 4% (v/v) paraformaldehyde fixation solution for 30 min at room temperature. Afterward, collect paraformaldehyde (*see* **Note 7**).
5. Wash wells with fixed matrices and cells once with PBS and then permeabilize the cells with 0.5% (v/v) Triton X-100 permeabilization buffer for 10 min at room temperature and stain with DAPI (1,10,000 in PBS) for 15 min at room temperature.
6. If cells were cultured on glass coverslips, these can be mounted as explained in **steps 3.4.8–12**.
7. Nuclei of cells growing on top of the matrices can be imaged with a 4x objective on a fluorescent microscope. It is recommended to take at least eight images per well. Images of DAPI-stained nuclei can be analyzed manually or by using an image analysis software such as FIJI (ImageJ, https://imagej.net/) or CellProfiler (https://cellprofiler.org/).

3.8 Optional: Colony Formation Assay on Top of Matrices

1. After preparing the matrices and cells (*see* **Note 17**) according to **step 3.5**, dilute cells appropriately.
2. In a 24-well format, gently pipette 50–100 cells/well in 1 mL cell culture medium on top of the fibroblast-derived matrix. Incubate for at least 7 days at 37 °C in a cell culture incubator (5% CO_2) until colonies are visible. Medium can be replenished after 4 days by aspirating 0.5 mL and replenishing it with 0.5 mL of fresh culture medium.
3. Gently remove the growth medium and wash once with 1 mL PBS.
4. To fix and stain the colonies, add 1 mL of crystal violet staining solution. Incubate for 30 min at room temperature.
5. Collect the crystal violet solution (*see* **Note 7**) and wash plates with copious amounts of water. Dry plates slightly tilted and upside down on a paper towel.
6. Image plates on a flatbed scanner and quantify the number of colonies manually or by using an image analysis software such as FIJI (ImageJ, https://imagej.net/) and the "Analyze Particles" function. Sample images are shown in Fig. 3a, b.

3.9 Optional: Tumor Cell Migration on Top of Matrices

1. To assess tumor cell migration on the decellularized extracellular matrices, spheroids are formed by seeding 2×10^3 tumor cells into the wells of an ultralow adherence 96-well U-bottom plate in 100 μL cell culture medium and are cultured for 72 h at 37 °C in a cell culture incubator (5% CO_2) to form spheroids (*see* **Note 22**).
2. Prepare the matrices according to **step 3.5**. Using a p1000 pipette, carefully remove the spheroids from the ultralow adherence 96-well U-bottom plate one at a time and place them on top of the matrices (*see* **Note 11**). Four spheroids can be added to one well when matrices are generated in a 24-well plate (*see* **Note 23**).
3. Outward migration of the tumor cells from the spheroids is imaged using a 4x or 10x objective on an automated live cell imager or on a standard (fluorescent) microscope at appropriate intervals.
4. The area covered by migrated cells can be quantified using the manual area measurement tool in FIJI (ImageJ, https://imagej.net/) or surface area applications in live cell imagers. Sample images are shown in Fig. 3c.

3.10 Optional: Tumor Cell Invasion/Transmigration Through Matrices

1. For transwell invasion assays through decellularized fibroblast-derived matrix, these need to be prepared in 8-μm pore size transwells according to Subheadings 3.1–3.3 and 3.5 omitting the glutaraldehyde step (*see* **Notes 1** and **9**).

2. Transwell inserts with deposited matrices on top are placed into a fresh 24-well plate and 500 μL cell culture medium with FBS is pipetted into the bottom chamber.
3. Seed 4×10^4 tumor cells in 200 μL serum-free medium into the top chamber of the transwell on top of the matrix (*see* **Note 11**). Incubate for 24–96 h at 37 °C in a cell culture incubator (5% CO_2) (*see* **Note 24**).
4. After cells have invaded through the matrix and to the bottom of the transwell, remove the medium from the transwells and wash both the top and bottom chambers with 0.5 mL PBS.
5. Cells that have not migrated through the membrane are removed from the top using a cotton bud.
6. For fixation and staining, transwell inserts are placed into 500 μL crystal violet solution. Incubate for 30 min at room temperature.
7. The stained transwell inserts are washed three times with 500 μL PBS (top and bottom) and then left to dry up-side-down on a filter paper with the bottom of the transwell insert (and the cells adhered to it) facing upward.
8. Transmigrated cells can be imaged by placing the transwell insert on a glass slide and using an inverted microscope (phase imaging for gray scale or RGB-mode illumination for color images) equipped with a 4x objective.
9. The percentage area of each transwell insert covered with transmigrated cells can be quantified in FIJI (ImageJ, https://imagej.net/) using the Analyze Particles function.

4 Notes

1. If the fibroblast-derived matrix is generated for imaging on an inverted fluorescent microscope, it can also be generated in a glass bottom plate. Of note, matrices generated on coverslips can be mounted on glass slides and stored for longer periods at 4 °C to −20 °C. If the matrix is not generated for fluorescence imaging, for example, for colony formation assays, alternative cell culture vessels, such as 6-well or 12-well dishes, can be used. Volumes throughout this protocol will need to be adjusted accordingly. For invasion assays, 8-μm pore transwells are used. Smaller culture vessels such as 96-well plates are not recommended.
2. Alternative media can be used as required for other fibroblast cell lines or primary fibroblasts.
3. While strictly not required, acid pre-treatment can improve ECM adsorbance and cell adherence.

4. Both Vectorshield mounting medium (without DAPI) and Prolong Gold can be used. The latter needs curing overnight.
5. Gelatin coating and cross-linking with glutaraldehyde are not strictly required. However, it is particularly recommended when using glass coverslips or glass plates, as coating and cross-linking will anchor matrices more strongly to the glass surface.
6. Glutaraldehyde cross-linking is not recommended when generating matrices inside of transwells for invasion assays.
7. Glutaraldehyde, paraformaldehyde, and crystal violet are toxic and must be disposed according to local regulations.
8. Coated coverslips/glass bottom plates/cell culture plates can be stored for future use. Add 1 mL sterile PBS and add 1% penicillin/streptomycin to each well, seal the plate with parafilm to prevent evaporation, and store at 4 °C for up to 1 week. Before using the coated plates/coverslips, aspirate PBS and continue with **step 3.2**.
9. Cell numbers need to be optimized to generate confluent cell layers. A starting point for 24-wells are $2.5–5 \times 10^4$ cells/mL, for 6-wells $1–2.5 \times 10^5$ cells/well, and for 24-well 8-μm transwell inserts 1×10^4 cells/transwell.
10. It is recommended to test different time points, particularly if fibroblasts are treated with growth factors or drugs. 3T3 and MRC5 fibroblasts will give sufficient matrix after 7 days, while primary fibroblasts are recommended to be cultured for 10–14 days and to receive daily medium changes.
11. As the generated matrices are very fragile, pipette tips should never touch the cell or matrix layer, and no vacuum aspirator should be used for any of the following steps.
12. This protocol primarily leads to the deposition of fibronectin; however, other matrix proteins such as collagens can also be detected and imaged. As an alternative to staining collagens with antibodies, second harmonic imaging can be used.
13. To use a smaller amount of antibody, a humidity chamber can be used. For this, place filter paper into a 15-cm culture plate and soak the filter paper with sterile PBS. Add a parafilm strip large enough for several coverslips on top. Per coverslip, add a 45 μL drop of primary antibody solution onto the parafilm. Use a needle with a bent tip as a grappling hook to help lift off the coverslips from the well bottoms while grabbing them with forceps to remove them from the wells. Dip the coverslips in dH_2O, and then gently tap the edges of the coverslips on a piece of filter paper to remove any excess liquid. Gently place the coverslips with the matrix-coated side facing downward onto the antibody drop. Close the lid of the plate to generate

a humidity chamber and incubate for 1 h at room temperature. After the primary antibody incubation, place the coverslips back into the 24-well plate for subsequent washing steps.

14. For the secondary antibody incubation, the humidity chamber can also be used. Use the same procedure as described in **Note 13**. After the incubation, place the coverslips back into the 24-well for the PBS washes.
15. DAPI (1:10,000 diluted in PBS) can be used instead of PBS alone to visualize insufficient cell lysis and any remaining nuclei. A further PBS wash without DAPI is recommended in this case.
16. Gentle but thorough washing is important since residual lysis buffer or DNase I can harm the cells that are re-seeded on top of the matrices.
17. Cells expressing a stable fluorescent protein can be used to monitor colony formation over time. Temporary labels that can be toxic over longer periods or get diluted in proliferating cells are not recommended for this assay.
18. Cell proliferation over time can be imaged using an automated live cell imager.
19. Washing with PBS++ containing magnesium and calcium is particularly important to sustain activated integrins and focal adhesions.
20. Alternatively, cells can be pre-labelled with Hoechst 33342, or cells expressing a stable fluorescent protein with a nuclear localization can be used to enable visualization of the live cells. This would also avoid the necessity to label cells post-fixation with DAPI.
21. Adhesion times may vary between different cell models, and time points can be shortened or prolonged accordingly.
22. Capability of tumor cells to form stable and solid spheroids should be tested prior to the start of this protocol. Spheroids need to be stable enough to be transferred with a pipette. Cells expressing a stable fluorescent protein or labelled with a non-toxic membrane stain (e.g., the CellTracker™ dye range) can be used to monitor cell migration.
23. When placing multiple spheroids into one well with the matrices, ensure that the spheroids do not cluster together. This can be achieved by releasing the spheroids above different parts of the matrix and by briefly waiting in-between, which allows the spheres to settle down. It is recommended to leave the plate for 30 min in the tissue culture cabinet before moving them back in to the incubator to allow the spheres to settle down. Be careful not to disturb the plate when taking it in and out of the incubator.

24. Cells might need longer to migrate through a matrix-coated transwell insert and invasion time will vary depending on the tumor cell line.

Acknowledgments

This work was supported in parts by grant 016/S/20 from the British Skin Foundation awarded to Ute Jungwirth, by grant BB/T012978/1 from the Biotechnology and Biological Sciences Research Council BBSRC/UKRI awarded to Gernot Walko, and by grant MR/W006308/1 for the GW4 BIOMED2 DTP awarded to the Universities of Bath, Bristol, Cardiff and Exeter from the Medical Research Council (MRC)/UKRI. Figure 1 was partly generated using Servier Medical Art, provided by Servier, licensed under a Creative Commons Attribution 3.0 unported license.

References

1. Lu P, Takai K, Weaver VM, Werb Z (2011) Extracellular matrix degradation and remodeling in development and disease. Cold Spring Harb Perspect Biol 3:a005058. https://doi.org/10.1101/cshperspect.a005058
2. Naba A, Pearce OMT, del Rosario A et al (2017) Characterization of the extracellular matrix of normal and diseased tissues using proteomics. J Proteome Res 16:3083–3091. https://doi.org/10.1021/acs.jproteome.7b00191
3. Caligiuri G, Tuveson DA (2023) Activated fibroblasts in cancer: perspectives and challenges. Cancer Cell 41:434–449. https://doi.org/10.1016/j.ccell.2023.02.015
4. Bartoschek M, Oskolkov N, Bocci M et al (2018) Spatially and functionally distinct subclasses of breast cancer-associated fibroblasts revealed by single cell RNA sequencing. Nat Commun 9:5150. https://doi.org/10.1038/s41467-018-07582-3
5. Jungwirth U, van Weverwijk A, Evans RJ et al (2021) Impairment of a distinct cancer-associated fibroblast population limits tumour growth and metastasis. Nat Commun 12:3516. https://doi.org/10.1038/s41467-021-23583-1
6. Naba A, Clauser KR, Ding H et al (2016) The extracellular matrix: tools and insights for the "omics" era. Matrix Biol 49:10–24. https://doi.org/10.1016/j.matbio.2015.06.003
7. Hata R, Senoo H (1989) L-ascorbic acid 2-phosphate stimulates collagen accumulation, cell proliferation, and formation of a three-dimensional tissuelike substance by skin fibroblasts. J Cell Physiol 138:8–16. https://doi.org/10.1002/jcp.1041380103
8. Matsuda S, Iwata H, Se N, Ikada Y (1999) Bioadhesion of gelatin films crosslinked with glutaraldehyde. J Biomed Mater Res 45:20–27. https://doi.org/10.1002/(sici)1097-4636(199904)45:1<20::aid-jbm3>3.0.co;2-6
9. Puspoki Z, Storath M, Sage D, Unser M (2016) Transforms and operators for directional bioimage analysis: a survey. Adv Anat Embryol Cell Biol 219:69–93. https://doi.org/10.1007/978-3-319-28549-8_3
10. Wershof E, Park D, Barry DJ et al (2021) A Fiji macro for quantifying pattern in extracellular matrix. Life Sci Alliance 4:e202000880. https://doi.org/10.26508/lsa.202000880

Chapter 3

Induced Formation of Plasma Membrane Protrusions with Porous Materials as Instructive Surfaces

Morteza Aramesh and Cecilia Persson

Abstract

The plasma membrane is a vital component in cellular processes, and its structure has a significant impact on cellular behavior. The physical characteristics of the extracellular environment, along with the presence of surface pores, can influence the formation of membrane protrusions. Nanoporous surfaces have demonstrated their capacity to induce membrane protrusions in both adherent and non-adherent cells. This chapter presents a methodology that utilizes a nanoporous substrate with nanotopographical constraints to effectively stimulate the formation of membrane protrusions in cells.

Key words Plasma membrane shape, Porous surfaces, Cell instructive materials, Microvilli and filopodia induction

1 Introduction

The plasma membrane is a component of cellular architecture, serving as a crucial boundary that separates the intracellular content from the extracellular environment. Its intricate shape and dynamics play a pivotal role in regulating various fundamental cellular processes, including cell division, migration, and intercellular signaling [1]. One of the factors influencing plasma membrane shape is the physical properties of the extracellular environment [2, 3]. For instance, the viscoelasticity and topography of the extracellular matrix can induce changes in plasma membrane shape, subsequently affecting cell behavior [4, 5]. Additionally, the presence of pores or cavities on a substrate's surface can influence plasma membrane morphology [6–9]. By precisely controlling

Christoph Wülfing and Robert F. Murphy (eds.), *Imaging Cell Signaling*, Methods in Molecular Biology, vol. 2800,
https://doi.org/10.1007/978-1-0716-3834-7_3,

the shape and size of these pores or cavities, one can create a physical environment that guides the formation of plasma membrane structures, such as filopodia, lamellipodia, or microvilli.

Protrusions on the cell membrane, such as microvilli and filopodia, play crucial roles in cell signaling. They enhance cell surface receptor localization and clustering, increasing the sensitivity and efficiency of receptor-mediated signaling. These protrusions provide additional surface area for the localization of receptors, facilitating the interaction between receptors and their ligands. By enhancing receptor clustering, protrusions optimize signal transduction processes and promote effective cell signaling.

Several studies have demonstrated the effectiveness of utilizing porous surfaces to control cellular behavior [6, 10]. Various techniques, including ion irradiation [11, 12], lithography [13], anodization [14–16], deposition [17], and particulate leaching [18, 19], can be employed to fabricate porous surfaces to instruct membrane protrusion formation. The emergence of 3D-printing has further expanded the range of possibilities for creating porous structures with multiscale properties, holding promising applications in bone generation and other implant-related fields [18, 19].

By precisely manipulating the size, shape, and distribution of pores or cavities on a surface, one can guide the formation of membrane protrusions, such as filopodia or microvilli in a controlled manner. This provides a valuable tool for investigating the impact of membrane protrusions on cellular processes like migration, adhesion, and signaling. For instance, when cells are cultured on a nanoporous surface with small pores ranging from 100 to 400 nm, they tend to develop microvilli [6]. Conversely, when cells are cultured on surfaces featuring larger pores in the micron range, it can lead to the formation of more substantial protrusions, such as filopodia [6]. The formation of membrane protrusions is believed to trigger biological processes, including the assembly of signaling microclusters [8]. This phenomenon holds particular significance in T cells, where membrane protrusions in the range of 100–1000 nm in diameter have shown to enhance antigen recognition by facilitating amplified and sustained signaling at their tips. The physical barrier created by nanopores, possibly resulting from the size-exclusion of membrane proteins or other cellular components into the small pores, along with the high membrane curvature at the edges of protrusions, is thought to contribute to this effect [6]. This chapter presents an approach to induce the formation of microvilli and filopodia in cells by utilizing substrates with nanotopographical features (Fig. 1).

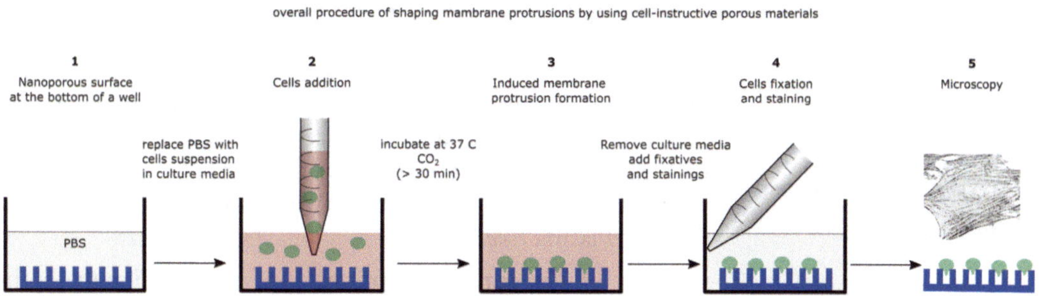

Fig. 1 The illustration of the step-by-step process involved in utilizing cell-instructive porous materials to shape membrane protrusions and microscope imaging

2 Materials

Prepare all solutions using phosphate buffer solution (PBS) 1× and pH 7.4.

2.1 Nanoporous Surfaces

1. Nanoporous surfaces can be obtained through various sources and procedures, including anodization, salt leaching, or ion-track etching (*see* **Note 1**).
2. The sample size can vary depending on the experimental setup, but for the purposes of the calculations outlined in this protocol, it is assumed that a 24-well plate is used, with each well having a diameter of approximately 16.7 mm.
3. Equipment: Basic Plasma Cleaner, PDC-32G from Harrick Plasma, USA.

2.2 Chemicals

1. The formation of protrusions inside nanopores can be achieved without the need for any chemical modifications (*see* **Note 2**).
2. The material should be sterile, and the surface of the pores should be hydrophilic (*see* **Note 3**).

2.3 Cells

Prepare the cells in a desired culture media with the density of 2.5×10^5 cells.

2.4 Immunostaining

1. Fixation buffer: 4% Paraformaldehyde (PFA) and 0.1% Triton-X in PBS.
2. Quenching buffer: 0.01% $NaCl_4$ in PBS.
3. Blocking buffer: 2% bovine serum albumin (BSA) in PBS.
4. Actin and nucleus staining buffer: Phalloidin (Alexa Fluor 488, 546, 647 from Thermo Fisher Scientific) in 1:200 dilution, DAPI (4′,6-Diamidino-2-Phenylindole, Dihydrochloride) in 5 μg/mL concentration.
5. Prolong gold antifade reagent.

3 Methods

3.1 Material Preparation

3.1.1 Surface Treatment (Optional)

1. Oxygen or air plasma effectively removes organic material from surfaces while simultaneously creating hydrophilic surfaces with enhanced wetting properties. This process can be achieved within a duration of 3 min at 18 W, utilizing a PDC-32G from Harrick Plasma (*see* **Note 4**).
2. Coating of the surfaces (optional): If desired, incubate the substrates with the proteins of interest (*see* **Note 2**).
3. Transfer the porous substrate to culture plates that are pre-filled with PBS 1x (*see* **Note 5**).

3.2 Cell Seeding

1. Remove the PBS and immediately pipette 0.5 mL of the cell's suspension per well in a 24-well plate containing the porous surfaces at the bottom.
2. Maintain the samples at 37 °C and a 5% CO_2 environment for the desired duration (*see* **Note 6**).

3.3 Fixation and Staining

1. Fixation buffer should be pre-warmed at 37 °C before start.
2. Gently remove the culture media from the wells (*see* **Note** 7).
3. Cover the samples with the fixation buffer for 10 min at room temperature.
4. Wash the samples with PBS three times (1 min, 1 min, 5 min).
5. (Optional) Incubate the samples with the quenching buffer for 10 min to reduce the autofluorescence and subsequently wash the samples with PBS (5 min).
6. (Optional) Incubate the samples with the blocking buffer for 1 h at room temperature before staining and subsequently wash the samples with PBS (5 min) to minimize nonspecific binding of antibodies or stains.
7. For actin and nucleus staining, incubate the samples with the staining buffer for 1 h at room temperature and dark.
8. For other (primary or secondary) antibodies, follow the manufacturer's protocol.
9. Wash the samples with PBS three times (1 min, 1 min, 5 min).
10. Mount the samples on a cover slip with an antifade reagent (*see* **Note 8**).
11. Visualize the samples with a confocal fluorescence microscope (*see* **Note 9**).
12. *See* **Note 10** for live cell imaging.

4 Notes

1. Some commercially available nanoporous surfaces include anodized aluminum oxide (AAO) membranes from different sources, such as Whatman Anodisc Membranes with pore sizes of 20 nm, 100 nm, and 200 nm, as well as ACS material AAO membranes with a range of pore sizes from 20 nm to 400 nm. Ion-etched polymer membranes, such as polystyrene and polycarbonate membranes with pore sizes of 200 nm and 400 nm, are also available from suppliers like Sterlitech. For in-house fabrication, nanoporous surfaces can be achieved through simple anodization processes using aluminum or titanium (Fig. 2). Please refer to the provided reference for more detailed information and fabrication protocols [20].
2. Alternatively, to explore biochemical interactions alongside protrusion formation, activating antibodies or adhesion molecules can be utilized. For instance, a streptavidin intermediate can be employed to coat the surface and facilitate the binding of biotinylated anti-CD3 and anti-CD28, which activates T cells. Another option is the use of ICAM-1 (Intercellular Adhesion Molecule 1) to cover the surface with adhesion molecules that bind to T cell integrins. For further details on surface functionalization with antibodies, please refer to Aramesh et al. [6].
3. For experiments involving cell culture over several days, autoclaving at 120 °C for 2 h is recommended. Alternatively, a 30 min treatment with 70% ethanol followed by three

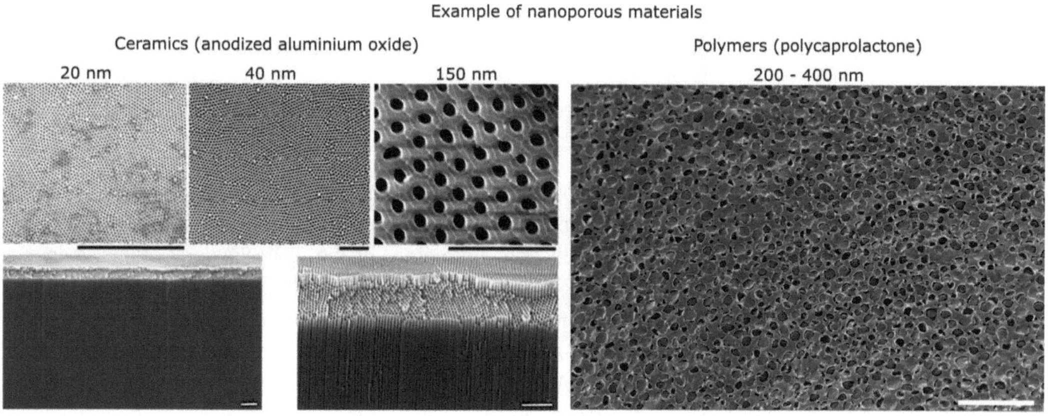

Fig. 2 Example of nanoporous structures. (left) SEM images of Anodic Aluminum Oxide (AAO) after the second anodization process are shown. The resulting porous structure exhibits a well-ordered arrangement with straight channels. The pore size of the membrane varies, depending on the specific anodization parameters employed. (right) SEM image of nanoporous polycaprolactone (PCL) scaffold. The scaffold was fabricated using porogens, which are sacrificial components that create voids within the PCL matrix. The resulting structure exhibits a highly porous structure with varying sizes. Scale bars indicate a length of 1 μm

consecutive washes with PBS can be performed to sterilize the surfaces. Note that not all materials are autoclavable.

4. Light-weighted samples can be carried away by the turbulent airflow created by the vacuum pump during the pumping/venting process of plasma cleaning. If possible, reduce the airflow during pumping and venting.
5. Pre-filing the wells with PBS simplifies the handling of the samples by reducing the electrostatic effects, making it easier to transfer them to the bottom of the wells without breaking.
6. Wait 60 min for the cells to form stable protrusions. For non-adherent cells, we recommend 30–60 min incubation. For adherent cells, the incubation time can be extended and may last up to several days.
7. During the fixation process, it is crucial to handle all pipetting steps with great care and gentleness. Approach the corner of the sample and delicately remove the culture media, ensuring it does not dry out. Similarly, gently pipette the buffers into the corner of the well, taking care to maintain a gentle and controlled flow (Fig. 3).

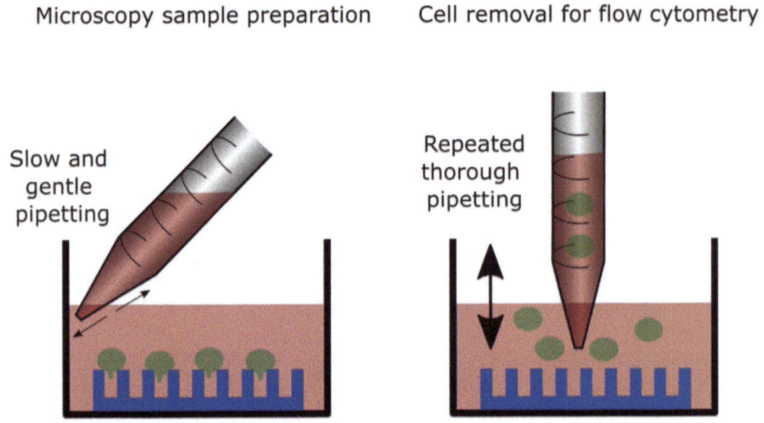

Fig. 3 Handling steps during fixation and cell detachment processes. Handle pipetting steps with care and gentleness during the fixation process. Delicately remove culture media from the corner of the well. Similarly, gently dispense buffers into the corner with a controlled flow. Cell detachment for flow cytometry: Detaching cells from the surface is essential for flow cytometry analysis. Trypsin is used for adherent cells, while extensive pipetting is required for non-adherent cells. Aspirating and expelling the medium repeatedly promotes detachment. On the other hand, for flow cytometry analysis, it is necessary to detach the cells from the surface. This can be achieved through the use of trypsin for adherent cells or by performing extensive pipetting for non-adherent cells (by repeatedly aspirating and expelling the medium, to agitate the cells and promote detachment

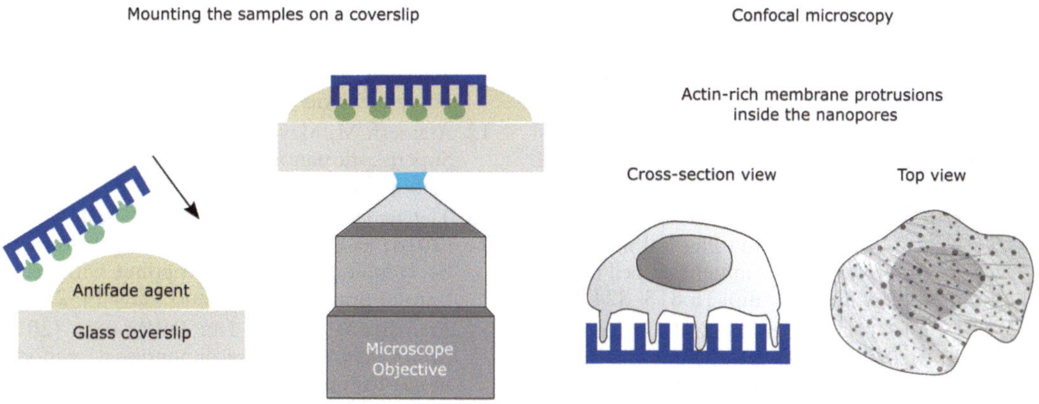

Fig. 4 Sample preparation for confocal optical microscopy: add mounting medium to a coverslip, flip the substrate onto the coverslip, and allow the mounting medium to cure before imaging. After identifying the dotted structures, adjust the microscope's focus to examine deeper into the sample. This allows for a rapid assessment of protrusion formation

8. After removing any excess solution, add 30 μL of Prolong Gold onto a H1.5 coverslip. Ensure that the cells are facing upward on the substrate, before flipping it on the coverslip with the mounting medium. To let the mounting medium cure, keep the samples in a dark environment for a minimum of 2 h. It is worth noting that at this stage, the samples can be stored in a 4 °C refrigerator for up to 5 days prior to imaging (Fig. 4).
9. Search for the presence of dotted structures (Fig. 4) and subsequently adjust the microscope's focus to examine deeper into the sample for a rapid assessment of protrusion formation. Conduct a z-scan using the lowest possible laser power to prevent fluorophore quenching. Begin the z-scan from the tip of the protrusions inside of the pores and conclude at the material's surface.
10. To visualize actin-rich protrusions in live cells, it is recommended to utilize stably transfected cells expressing fluorescent actin markers, such as Lifeact-GFP. The use of stably transfected cells with GFP-expressing actin yields optimal results. Transient actin polymerization-based dyes, on the other hand, are not efficient for live cell imaging inside nanopores.

Acknowledgments

This work is conducted within the Additive Manufacturing for the Life Sciences Competence Center (AM4Life). The authors gratefully acknowledge financial support from Sweden's Innovation Agency VINNOVA (Grant no: 2019-00029). MA acknowledges Prof. Viola Vogel for the mentorship and support to explore cell-surface interactions.

References

1. Schmick M, Bastiaens PI (2014) The interdependence of membrane shape and cellular signal processing. Cell 156:1132–1138
2. Ammendolia DA, Bement WM, Brumell JH (2021) Plasma membrane integrity: implications for health and disease. BMC Biol 19:1–29
3. Charras G, Sahai E (2014) Physical influences of the extracellular environment on cell migration. Nat Rev Mol Cell Biol 15:813–824
4. Chaudhuri O, Cooper-White J, Janmey PA et al (2020) Effects of extracellular matrix viscoelasticity on cellular behaviour. Nature 584: 535–546
5. Jain N, Vogel V (2018) Spatial confinement downsizes the inflammatory response of macrophages. Nat Mater 17:1134–1144
6. Aramesh M, Stoycheva D, Sandu I et al (2021) Nanoconfinement of microvilli alters gene expression and boosts T cell activation. Proc Natl Acad Sci USA 118:e2107535118
7. Ariga K (2022) Mechano-nanoarchitectonics: design and function. Small Methods 6: e2101577
8. Sánchez MF, Tampé R (2023) Ligand-independent receptor clustering modulates transmembrane signaling: a new paradigm. Trends Biochem Sci 48:156–171
9. Hyun J, Kim SJ, Cho S-D et al (2023) Mechano-modulation of T cells for cancer immunotherapy. Biomaterials 297:122101
10. Xia J, Yuan Y, Wu H et al (2020) Decoupling the effects of nanopore size and surface roughness on the attachment, spreading and differentiation of bone marrow-derived stem cells. Biomaterials 248:120014
11. Dutta S, Hussain CM (2022) Membranes with functionalized nanomaterials: current and emerging research trends in membrane technology. Elsevier, Amsterdam
12. Aramesh M, Mayamei Y, Wolff A et al (2018) Superplastic nanoscale pore shaping by ion irradiation. Nat Commun 9:835
13. Tiefenauer RF, Tybrandt K, Aramesh M et al (2018) Fast and versatile multiscale patterning by combining template-stripping with nanotransfer printing. ACS Nano 12:2514–2520
14. Aramesh M, Fox K, Lau DWM et al (2014) Multifunctional three-dimensional nanodiamond-nanoporous alumina nanoarchitectures. Carbon 75:452–464
15. Aramesh M, Cervenka J (2014) Surface modification of porous anodic alumina for medical and biological applications. In: Seifalian A, del Mel A, Kalaskar DM (eds) Nanomedicine. One Central Press, Manchester
16. Losic D, Santos A (2015) Nanoporous alumina: fabrication, structure, properties and applications. Springer, New York
17. Aramesh M, Djalalian-Assl A, Yajadda MMA et al (2016) Thin nanoporous metal-insulator-metal membranes. ACS Appl Mater Interfaces 8:4292–4297
18. Jakus AE, Geisendorfer NR, Lewis PL et al (2018) 3D-printing porosity: a new approach to creating elevated porosity materials and structures. Acta Biomater 72:94–109
19. Yassin MA, Fuoco T, Mohamed-Ahmed S et al (2019) 3D and porous RGDC-functionalized polyester-based scaffolds as a niche to induce osteogenic differentiation of human bone marrow stem cells. Macromol Biosci 19:e1900049
20. Lee W, Park S-J (2014) Porous anodic aluminum oxide: anodization and templated synthesis of functional nanostructures. Chem Rev 114:7487–7556

Chapter 4

Design and Synthesis of DNA Origami Nanostructures to Control TNF Receptor Activation

Göktuğ Aba, Ferenc A. Scheeren, and Thomas H. Sharp

Abstract

Clustering of type II tumor necrosis factor (TNF) receptors (TNFRs) is essential for their activation, yet currently available drugs fail to activate signaling. Some strategies aim to cluster TNFR by using multivalent streptavidin or scaffolds based on dextran or graphene. However, these strategies do not allow for control of the valency or spatial organization of the ligands, and consequently control of the TNFR activation is not optimal. DNA origami nanostructures allow nanometer-precise control of the spatial organization of molecules and complexes, with defined spacing, number and valency. Here, we demonstrate the design and characterization of a DNA origami nanostructure that can be decorated with engineered single-chain TNF-related apoptosis-inducing ligand (SC-TRAIL) complexes, which show increased cell killing compared to SC-TRAIL alone on Jurkat cells. The information in this chapter can be used as a basis to decorate DNA origami nanostructures with various proteins, complexes, or other biomolecules.

Key words DNA nanotechnology, Tumor necrosis factor receptors, TNF-related apoptosis-inducing ligand (TRAIL), Death receptor 5 (DR5), Receptor clustering

1 Introduction

Cells sense extracellular signals and change their cellular functions in response [1, 2]. Oftentimes, selective receptor-ligand interactions are the initial step for cellular activation [1, 3]. In particular, clustering of cell surface receptors is essential for activation of tumor necrosis factor (TNF) receptor superfamily (TNFRSF), which has important roles in mammalian cell proliferation, cell death, immune regulation, and morphogenesis [4–6]. The spatial organization of the cell surface receptors can be crucial for the regulation of cellular signaling cascades. This has been described for a number of receptor-ligand interactions, for example, TNF-related apoptosis-inducing ligand (TRAIL) and death receptor 5 (DR5) [1, 7–9].

Despite the current knowledge about the structure of the TNFRSF receptors (TNFRs), currently available drugs fail to

Christoph Wülfing and Robert F. Murphy (eds.), *Imaging Cell Signaling*, Methods in Molecular Biology, vol. 2800,
https://doi.org/10.1007/978-1-0716-3834-7_4,

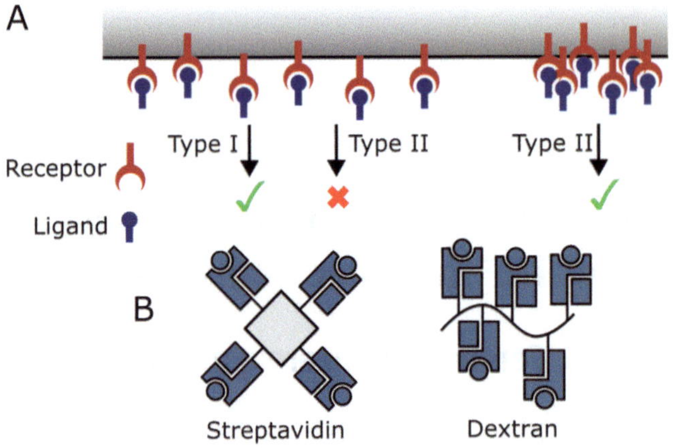

Fig. 1 Receptor clustering leads to activation of TNFR. (**a**) Schematic overview of type I and type II TNFRs. Type II TNFRs require receptor clustering for activation of signaling, unlike type I TNFRs, which can be activated by the interaction between the ligand and the receptor [11, 12]. (**b**) Schematic representation of two different multimerization techniques of major histocompatibility complexes (MHCs). MHCs are tetramerized using streptavidin on the left and multimerized on a dextran backbone on the right [20, 21]

comprehensively activate signaling in type II TNFRs, such as 4-1BB, CD40, and death receptor 5 (DR5) [10, 11]. Interestingly, efficient receptor activation was shown when their ligands were presented in membrane-associated form (Fig. 1a) [11, 12]. Many agonistic antibodies require cross-linking for effective induction of the receptors [4, 13]. Furthermore, the interligand distance affects signaling activation in immunological synapses [14, 15]. Some strategies multimerize ligands using multivalent streptavidin, or scaffolds based on peptides, dextran, or graphene, to successfully initiate DR5-mediated cell signaling (Fig. 1b) [16–21]. However, these methods do not allow precise control over the interligand distance, and achieving a ligand arrangement with sub-10 nm distancing is challenging [15, 22].

In comparison, DNA origami nanostructures [15, 23, 24] are an ideal alternative; DNA origami nanostructures allow nanometer-precise control over the spatial organization of ligands, with defined composition and number (Fig. 2) [25]. Moreover, DNA nanostructures are stable in the presence of human sera [26]. DNA nanostructures have been previously used to activate Toll-like receptors using CpG-motifs [25], DR5 using TRAIL-like peptides [15], and T cell antigen receptors (TCRs) by peptide/major histocompatibility complexes (pMHCs) [27]. We will not cover all terminology of DNA nanotechnology, but several excellent reviews are available as resources (see e.g., [28]).

There are few agonistic antibodies in preclinical development or in the clinic that can induce receptor super-clustering. Receptor

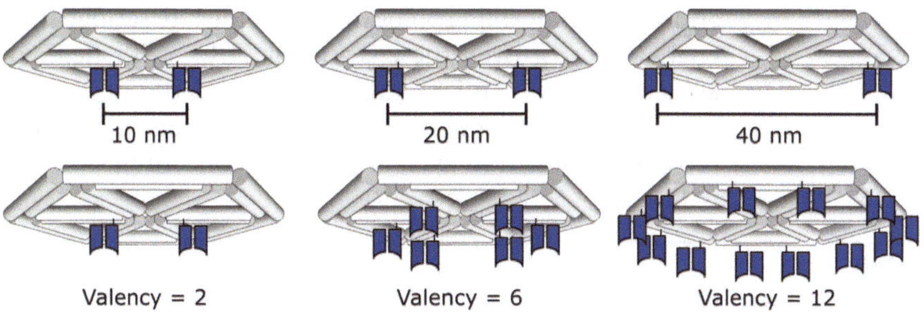

Fig. 2 DNA origami nanostructures allow great control over the spatial organization of ligands with nanometer precision and defined composition and number. Double-stranded DNA helices are shown as gray cylinders. Here, ligands are represented as Fab fragments, shown in dark blue

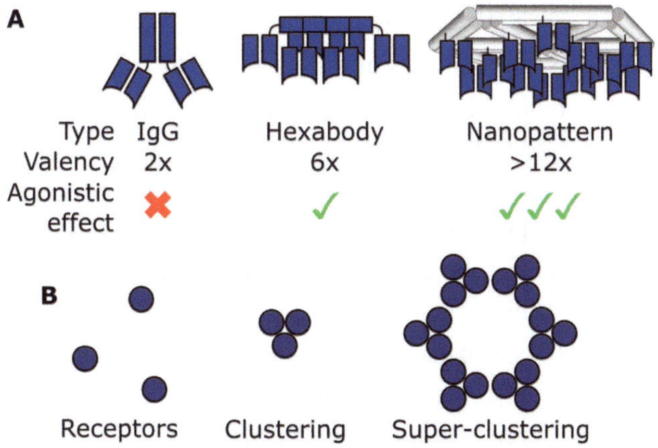

Fig. 3 Clustering of receptors improves downstream agonistic stimulation. (**a**) DNA nanopatterned antibody/ligand clustering can provide higher binding due to higher avidity, potentially increasing the agonistic effects compared to other super-clustering methods. (**b**) Schematic representation of receptor clustering and super-clustering

clustering can be defined as bringing multiple receptors close together on a 2D membrane to enable activation, while receptor super-clustering brings preformed receptor clusters together, which can increase the agonistic effects even further (Fig. 3). Similar to other TNFRSF ligands (TNFL), TRAIL forms homomeric trimers [11]. TNFL trimers bind to their corresponding monomeric receptor on a cell surface to initiate trimerization of the receptors to activate downstream signaling pathways [6]. In this context, trimerization of the receptors can be considered as the receptor clustering. Proteins can be engineered to assist receptor clustering, for example, single-chain mutant variants of TRAIL (SC-TRAIL) were engineered to improve the agonistic effects [29]. This was achieved by linking three extracellular domains of TRAIL by peptide linkers. Here, we combine protein engineering of SC-TRAIL with DNA nanotechnology to super-cluster DR5 and assess cell killing. The methodology described herein could enable control over the spacing of DR5, with defined valencies and geometries, and the ability to study the optimal intraligand distancing. Knowledge of these parameters may enable strong receptor stimulation even at very low concentrations.

2 Materials

2.1 Reagents and Materials

1. M13mp18 genome, Tebubio
2. Amicon Ultra 0.5–100 kDa filters (Millipore)
3. Amicon Ultra-15 50 kDa concentrator (Millipore)
4. Glow-discharged C-flat carbon film 200 mesh grids (Aurion, Netherlands)
5. 2% aqueous uranyl formate (SPI-Chem, 02545-AA)
6. pCPF3.05 vector: a gift from Robbert Kim (Protein Facility, LUMC)
7. *E. coli* Rosetta2 (DE3) BL21 derivative (Novagen)
8. HisPur™ Ni-NTA Resin (Thermo Fisher)
9. Bolt™ 4–12%, Bis-Tris, 1.0-mm SDS-PAGE gels (Thermo Fisher)
10. SimplyBlue™ SafeStain (Thermo Fisher)
11. HiLoad® 16/600 Superdex® 75 size exclusion chromatography column (Merck)
12. Dibenzocyclooctyne-N-hydroxysuccinimidyl ester (Sigma Aldrich, 761524)
13. Amino-modified DNA handles (IDT)
14. Jurkat T cells: a kind gift from Mirjam Heemskerk

2.2 Buffers

1. Folding buffer: 20-mM Tris, 5-mM NaCl and a range of $MgCl_2$ concentrations, between 12 and 40 mM, to determine the optimum $MgCl_2$ concentration.
2. Nickel binding buffer: 50-mM Tris pH 7.5, 150-mM NaCl, and 100-μM $ZnCl_2$.
3. Wash buffer: 50-mM Tris pH 7.5, 150-mM NaCl, 5-mM $MgCl_2$, 100-μM $ZnCl_2$, 10-mM imidazole, and 10% glycerol.
4. Elution buffer: 50-mM Tris pH 7.5, 150-mM NaCl, 100-μM $ZnCl_2$, 250-mM imidazole, and 10% glycerol.
5. SC-TRAIL stock buffer: 50-mM Tris pH 7.5, 150-mM NaCl, 100-μM $ZnCl_2$, and 10% glycerol.
6. Sortase buffer: 50-mM Tris pH 7.5, 150-mM NaCl, 10-mM $CaCl_2$, and 10% glycerol.

2.3 Equipment

1. Tecnai T12 Biotwin transmission electron microscope
2. Sonics Vibra-Cellä Ultrasonic Liquid Processor VCX 130
3. ÄKTA™ pure micro FPLC system (Cytiva)
4. TC20 Automated Cell Counter

2.4 Software

1. University of California, San Francisco (UCSF) Chimera: https://www.cgl.ucsf.edu/chimera/download.html
2. UCSF ChimeraX: https://www.cgl.ucsf.edu/chimerax/download.html
3. Cadnano2: https://cadnano.org/

3 Methods

3.1 Determining Protein-Protein Distances to Enforce Using DNA Nanotechnology

1. We manually checked the crystallographic packing of a range of TNFL/TNFR complexes to identify those closely packed in native-like conformations (i.e., all extracellular domains oriented parallel to each other). We identified a structure of a 1BBL/4-1BB complex that conformed to these constraints.
2. We hypothesized that the structurally homologous TNFL/TNFR family will all allow a similar packing regime, that this is the closest packing that a TNFL/TNFR array could achieve on a cell surface, and that the highest ligand/receptor density may also have the maximum receptor activation.
3. The crystallography data of the 4-1BBL/4-1BB complex can be visualized with UCSF ChimeraX [30] using the command "open 6MGP," which downloads and opens a protein database (pdb) file of a trimer of 4-1BBL molecules, each bound to a 4-1BB receptor (Fig. 4a) [31].
4. Typing the command "unitcell" shows a small region of the crystal packing of these trimers, and within the eight copies of the crystallographic unit cell created, there exists a pair of TNF/TNFR trimers closely packed (Fig. 4b; more easily visible with command "del #1 #2.1-4 #2.6 #2.8").
5. You can see the approximate distance between two of the trimers with the command "sel #2.5/C:242@CB #2.7/C:242@CB ; distance sel," which gives a value of ~5.5 nm.
6. This dimer of trimers can be used to create a hexameric platform, by duplicating the .pdb file and repeating the distance and angle of the dimers using the "matchmaker" command (Fig. 4c).
7. We can take this distance and hexagonal lattice and apply it to DR5/TRAIL to check that there are no steric clashes. We next aligned a model of DR5/TRAIL (pdb code 1D0G [32]) to this hexamer using the matchmaker command, and this model was saved in .pdb format for later use.

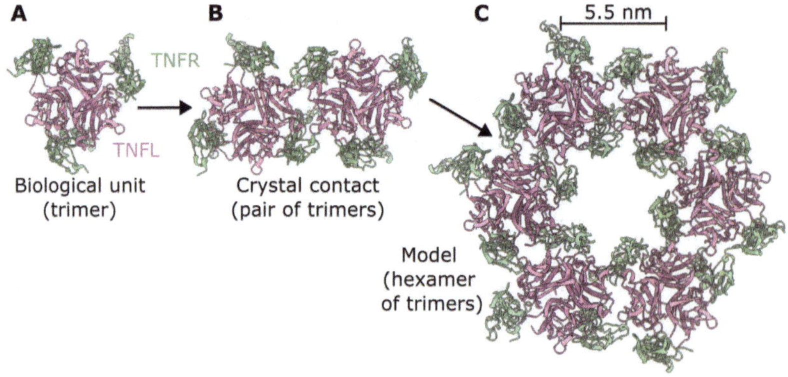

Fig. 4 Building a hexameric TNFR model. (**a**) Representation of the determination of the intraligand distancing between two different TNFL (pink) trimers when bound to their cognate TNFR complexes (green). Also shown are a dimer of trimers forming crystal contacts (**b**) and a theoretical model of a hexameric DR5/TRAIL complex (**c**)

3.2 Design a DNA Nanostructure

1. The DNA origami nanostructure was designed by writing a .ply file containing the geometry of our DNA design. The files can be created in a number of ways, see refs [33–35]. Below is the ply file for the hexagonal-base pyramid design we use, and the visualization of the ply file is shown in Fig. 5a.

```
ply
format ascii 1.0
element vertex 8
property float32 x
property float32 y
property float32 z
element face 12
property list uint8 int32 vertex_indices
end_header
0 0 0
-1 0 0
-0.5 0.866025403784439 0
0.5 0.866025403784439 0
1 0 0
0.5 -0.866025403784439 0
-0.5 -0.866025403784439 0
0 0 1.5
3 0 1 2
3 0 2 3
3 0 3 4
3 0 4 5
3 0 5 6
3 0 6 1
```

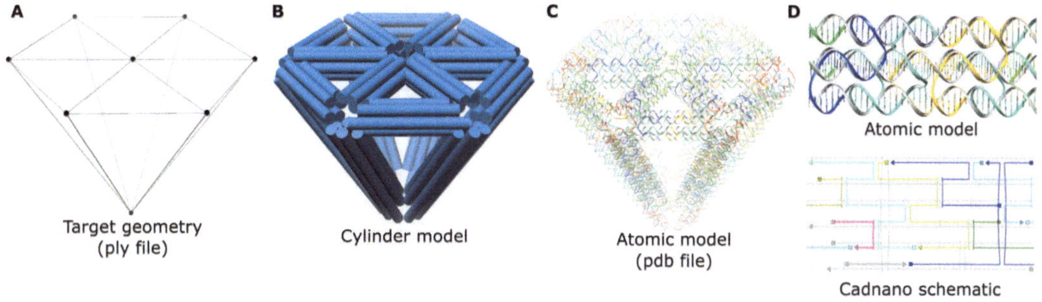

Fig. 5 Creation and modification of the DNA nanostructure using Athena

```
3 7 2 1
3 7 3 2
3 7 4 3
3 7 5 4
3 7 6 5
3 7 1 6
```

2. The ply file can be loaded into Athena [33] or alternatively can be uploaded onto one of the servers from the Bathe lab [34, 36, 37] to create all the required files.
3. In Athena, select Talos to create nanostructures with six DNA helices per edge (Fig. 5b), which are more rigid than the alternative Daedalus designs that use only two DNA helices per edge length [35].
4. Set the desired side length; 42 base pairs in this case, and then "Run Sequence Design." Next, select "Include PDB" and press Save. Athena will create all the files required to modify and synthesize a DNA nanostructure with the geometry provided in the ply file, including an atomic model (in .pdb format; Fig. 5c) and an associated cadnano file (in json format; Fig. 5d)

3.3 Identify Locations on the DNA Nanostructure to Correctly Place Ligands

1. Specific staple strands within the DNA nanostructure will be modified to bind to proteins; it is the location of these staple strands within our DNA nanostructure that will determine the spacing of the proteins. To determine which staples to modify, the atomic model obtained from Athena [33] was opened in UCSF Chimera [38], together with the TRAIL/DR5 hexamer model that we created earlier in step 1. Figure 6a shows a simplified model with white and blue representing the scaffold and staple strands of the DNA origami nanostructure, respectively, and pink the TRAIL proteins. The two structures have been moved with respect to one another such that the proteins are all above DNA helices.

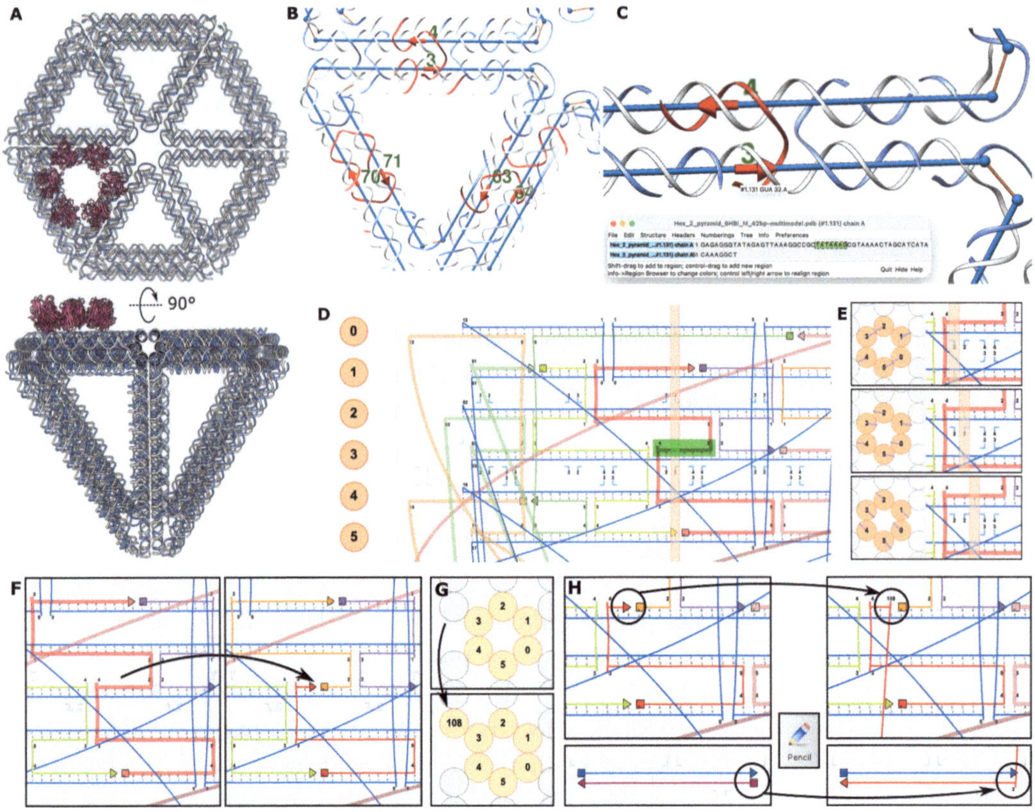

Fig. 6 Workflow of modifying the staple DNAs overlapping with the hexamer-of-trimers TRAIL model. (**a**) Models of the DNA nanostructure and hexameric TRAIL complex aligned so that each trimer is above a DNA helix. (**b**) Strands in locations that could be modified are colored red, and then the protein model is hidden for convenience. (**c**) The ID of the helices is determined in UCSF Chimera and the sequence shown. (**d**) The ID of the helices is shown in Cadnano, and the sequences matched to those of the pdb model to confirm the correct position is identified. (**e**) The pitch of the DNA helices is visible as blue lines within the orange circles, which represent DNA helices. As the orange bar is moved right, the blue line rotates clockwise, indicating the position of the scaffold strand. The staples are oriented directly opposite this blue line. The middle panel shows the situation where the staple strand is pointing directly out from the DNA nanostructure. (**f**) At this location, the staple can be broken in order to modify it. (**g**) An additional empty DNA helix is added at a convenient location (it does not need to be in any particular location; here it is shown adjacent to strand 3 for convenience). (**h**) A new DNA helix of the desired length is added to the new empty helix, before being connected to the previously identified location. Adding the DNA handle is now complete

2. In Fig. 6b, the staple strands close to the termini of the TRAIL proteins have been colored red; these are the staples that can be modified to present the proteins on a hexagonal array with a distancing of ~5.5 nm.

3. Also open the *guide_caDNAno.bild file output by Athena in UCSF Chimera (Fig. 6b) (*see* **Note 1**). The numbers of the DNA helices that contain the red staple strands can be seen, and these correspond to the numbers of the helices in the cadnano.json file (explained below).

4. The specific model IDs for the red staples were noted with the corresponding DNA sequence, which can be obtained with the sequence tool in UCSF Chimera (Tools > Sequence > Show Sequence Viewer) (Fig. 6c).

3.4 Identify Locations in the Cadnano File to Correctly Place Ligands

1. Open the.json file in Cadnano [39], where each double-stranded DNA helix is numbered (represented by the orange circles in Fig. 6d), and these correspond to the numbering in the "*guide_caDNAno.bild" file output from Athena.
2. The sequence of the scaffold strand can be added in Cadnano to further aid identification of each staple strand.
3. By comparing these files—the pdb model, the *guide_caDNAno.bild file, and the Cadnano [39] file—the locations identified in UCSF Chimera can be identified to determine which staples to modify. For example, in Fig. 6c and d, the same sequence and location has been highlighted in green.
4. In Fig. 6e, the orange circles represent the DNA helices. Within the orange circles are blue lines, which show the pitch of the scaffold DNA at the position shown by the orange rectangle on top of the DNA map. As the rectangle is moved right, the blue lines rotate clockwise (DNA is right-handed). Importantly, the staple DNA strands are at the opposite side to the scaffold strand and therefore opposite to the blue lines.
5. We want our strands to be modified pointing away from the DNA nanostructure. Therefore, to modify helix 3, we have chosen the location of the middle panel in Fig. 6e to include an extension to the staple, where the blue line points directly *in* to the DNA nanostructure and the staple therefore points directly *out* of the nanostructure (this location can be confirmed by comparing to the pdb model), and will therefore correctly display the protein. Thus, we will break the staple at the indicated position (Fig. 6f) (*see* **Note 2**).

3.5 Modify Staple Strands on the DNA Nanostructure for Ligand Binding

1. After locating the positions of the modifications, we need to extend the strands by 18–20 nucleotides. To do this, create an empty DNA helix by placing a new circle next to the strand you wish to modify (Fig. 6g). In this case, the strand is number 108, but this depends on the number of helices in your DNA nanostructure.
2. In the row for strand 108, draw a new scaffold/staple pair of the same length as the extension, in this case 19 nucleotides long (Fig. 6h).
3. Finally, using the pencil tool, connect the relevant termini of the staple strand and the extension (Fig. 6h). The desired sequence of the extensions can also be added at this point.

4. In this example, we have indicated six positions to modify the staple strands, but this DNA nanostructure can accommodate 24 modifications on a 5.5-nm lattice.
5. When all extensions have been added at the desired locations, it is helpful to color all the staples as groups. We usually use at least two colors; "core" staples—those that form the main structure—and "modification" staples, those that will display the proteins.
6. The sequence file can be exported from Cadnano as a .csv file. There will be numerous question marks "?" in the sequence file; these correspond to the unpaired single-stranded DNA staples that link the structure together [34]. These can all be replaced as Thymine (T) bases. In our case, all of the staples were modified to not exceed the 60-nucleotide limit including the extension sequence, as this saves a significant amount of money when ordering the staples.
7. The staple sequences were then ordered from Integrated DNA Technologies (IDT), already dissolved in water at a concentration of 100 μM, to be used for the folding reactions.

3.6 Design, Folding, and Purification of DNA Origami Nanostructures

1. After receiving the staples from IDT, the staples were thawed overnight at 4 °C.
2. The single-stranded viral M13mp18 genome was used as the scaffold. The final concentration of the scaffold DNA and each of the oligonucleotide strands from IDT were 20 nM and 200 nM in the folding reactions mixture, respectively.
3. After mixing together all required components in folding buffer, the folding reactions were placed in a Bio-Rad C1000 Touch Thermal Cycler and thermally annealed. To do so, the mixtures were held at 80 °C for 1 min, followed by a thermal annealing ramp from 80 °C to 75 °C (at a rate of 0.2 °C/min) and then subsequently from 75 °C to 30 °C (0.1 °C/min) and finally from 30 °C to 20 °C (0.1 °C/min). Figure 7a shows the agarose gel analysis of this folding reaction. The small shift in bands represents the folding of the scaffold DNA. The intensity of the bands represents the amount of folded DNA nanostructures.
4. It can be observed that the intensity of the bands decreases for $MgCl_2$ concentrations higher than 28 mM, suggesting that the folding reaction efficiency decreases above this concentration of $MgCl_2$. The excess staple DNA pool is also visible.
5. After determining that ~20 mM $MgCl_2$ was optimal, we then prepared one more folding reaction, this time with a range of NaCl concentrations, from 0 to 320 mM, to determine the optimum salt concentration (data not shown) [28].

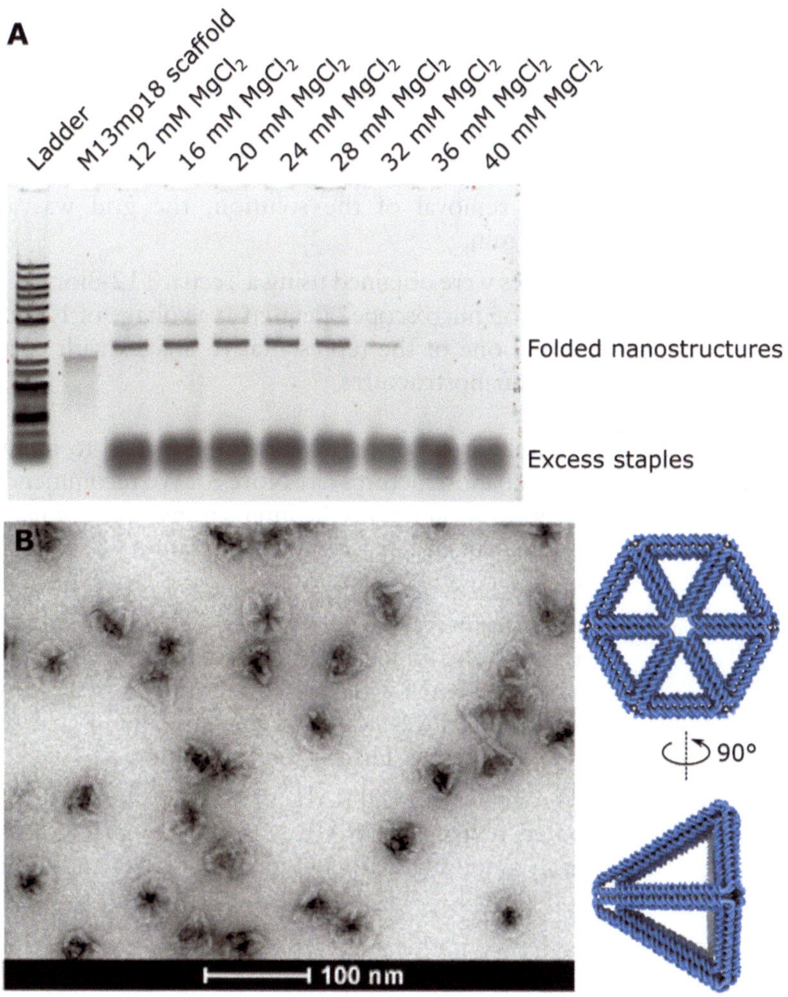

Fig. 7 Folding and characterization of DNA origami nanostructure. (**a**) Agarose gel analysis of the folding reactions with various $MgCl_2$ concentrations to determine the optimum salt concentrations. The locations of both excess staples and folded DNA origami nanostructures are indicated. 1% agarose gel was run at 100 V for 1 h with 3 μL samples that were diluted 12.5× in MQ. (**b**) Schematic representation of the designed DNA origami nanostructure with the corresponding transmission electron microscopy images obtained with negative staining. Scale bar: 100 nm

6. The folded structures were purified from excess oligonucleotides using the Amicon Ultra 0.5–100 kDa filters by washing five times with the folding buffer containing the determined optimum salt concentrations.

3.7 Electron Microscopy of Folded DNA Nanostructures

1. We prepared negative staining grids of the folding reaction with 20-mM $MgCl_2$ to inspect the folded nanostructures with transmission electron microscopy (TEM).

2. To do so, 10 μL of purified or unpurified samples were applied at 2 nM to glow-discharged C-flat carbon film 200 mesh grids and incubated for 1 min.
3. After removal of the excess solution, staining was performed using 10 μL of 2% uranyl formate and incubated for 10 s.
4. After removal of the solution, the grid was air dried for 5–15 min.
5. Images were obtained using a Tecnai T12 Biotwin transmission electron microscope operated at a voltage of 120 kV. Figure 7b shows one of the representative images with correctly folded DNA nanostructures.

3.8 Protein Design to Covalently Link SC-TRAIL to DNA

1. Various drugs or materials can be loaded to functionalize the DNA nanostructures (*see* **Note 3**). We recommend using DNA hybridization or specific interactions between nanobodies and their small targets for functionalization [40].
2. Specifically, we use a sortase enzyme [41] to covalently link proteins directly to DNA oligonucleotides and then hybridize these to the folded DNA nanostructure. This requires expressing SC-TRAIL with a with C-terminal sortag—a short peptide sequence LPXTG (where X is any amino acid, we use E)—followed by a 6-His tag for purification.
3. The gene for SC-TRAIL therefore comprised three concatenated repeats of TRAIL with 3× gly-gly-ser linkers between each monomer, with a C-terminal sortag and 6-His tag:

```
MGPQRVAAHITGTRGRSNTLSSPNSKNEKALGRKINS-
WESSRSGHSFLSNLHLRNGELVIHEKGFYYIYSQTYFRFQEEI-
KENTKNDKQMVQYIYKYTSYPDPILLMKSARNSCWSKDAEYGLY-
SIYQGGIFELKENDRIFVSVTNEHLIDMDHEASFF-
GAFLVGGGSGGSGGSPQRVAAHITGTRGRSNTLSSPNSKNEKALGR-
KINSWESSRSGHSFLSNLHLRNGELVIHEKGFYYIYSQTYFRFQEEI-
KENTKNDKQMVQYIYKYTSYPDPILLMKSARNSCWSKDAEYGLY-
SIYQGGIFELKENDRIFVSVTNEHLIDMDHEASFF-
GAFLVGGGSGGSGGSPQRVAAHITGTRGRSNTLSSPNSKNEKALGR-
KINSWESSRSGHSFLSNLHLRNGELVIHEKGFYYIYSQTYFRFQEEI-
KENTKNDKQMVQYIYKYTSYPDPILLMKSARNSCWSKDAEYGLY-
SIYQGGIFELKENDRIFVSVTNEHLIDMDHEASFF-
GAFLVGGGSGGSGGSLPETGGHHHHHH
```

4. This gene was ordered from Invitrogen GeneArt and cloned into the pCPF3.05 vector.

3.9 Expressing Sortagged SC-TRAIL

1. *E. coli* Rosetta2 (DE3) with pCPF3.05 encoding for SC-TRAIL was first inoculated in 100 mL Erlenmeyer flask with 10 mL LB-medium containing 50 μg/mL kanamycin in a shaker at 37 °C, 200 rpm for 17–18 h.
2. The culture broth was then inoculated in two 2-L Erlenmeyer containing 500 mL LB-medium with 50 μg/mL kanamycin until an OD 600 value of 0.5–0.6 was reached. 1-mM IPTG was applied to initiate the production of SC-TRAIL.
3. The culture broth was sampled and centrifuged at 6000 g for 1 min, and the pellet was dissolved in 50 μL Laemmli sample buffer (50-mM DTT) before addition of IPTG to monitor the SC-TRAIL production at the starting point.
4. Incubation took place at 18 °C 200 rpm for about 24 h, after which the culture broth was sampled and centrifuged again as described above. The culture was then harvested by centrifugation at 4000 g, 4 °C for 20 min, and the pellet was washed with 10 mL nickel binding buffer and transferred to a 50 mL tube.
5. The ~2.5 g pellet was dissolved in 10 mL wash buffer. The solution was afterward sonicated with Sonics Vibra-Cellä Ultrasonic Liquid Processor VCX 130 (total time: 1 min 45 s, pulse on: 15 s, pulse off: 45 s, amplitude: 50%).
6. The supernatant was collected after centrifugation (20 min at 16,000 g, 4 °C).
7. Protein purification was performed with HisPur™ Ni-NTA Resin.
8. The protein was eluted with different concentrations of imidazole containing elution buffer ranging from 50 mM to 250 mM imidazole.
9. Samples of the elutions were studied on a Bolt™ 4–12%, Bis-Tris, 1.0-mm SDS-PAGE gels with a running scheme of 200 V for 35 min.
10. Proteins were visualized by staining with SimplyBlue™ SafeStain.
11. SC-TRAIL containing samples were further purified and buffer exchanged to SC-TRAIL stock buffer using ÄKTA™ pure micro with HiLoad® 16/600 Superdex® 75 pg with 1 ml/min.
12. Elutions containing the SC-TRAIL were then concentrated using 15 mL Amicon Ultra-15 50 kDa (10 min, 3000 g, 4 °C).

3.10 Sortase-Mediated Transpeptidation Reaction

1. A small click handle was attached to the SC-TRAIL using the transpeptidation reaction mediated by the sortase enzyme (Fig. 8, reaction *i*). This was followed by the click reaction between the azide-containing SC-TRAIL and DBCO

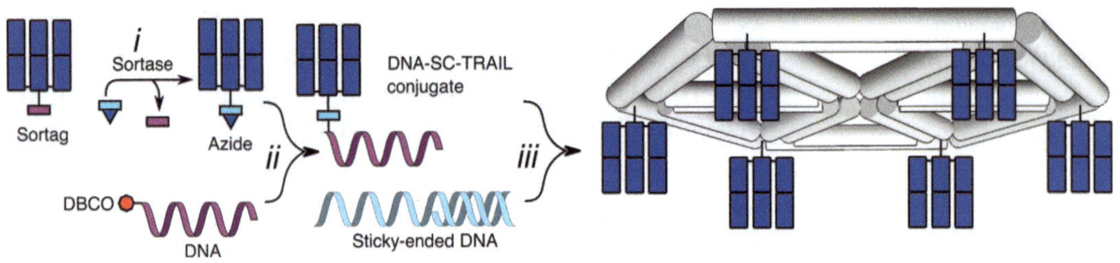

Fig. 8 Scheme for incorporating SC-TRAILs into DNA nano-structures using sortase reactions and click chemistry. Reaction i describes the usage of sortase enzyme for transpeptidation reaction for addition of azide to the protein. Reaction ii describes the click-reaction for covalently attachment of the DNA handle to the SC-TRAILs. Reaction iii shows the usage of DNA hybridization for the functionalization of DNA nanostructures

containing DNA handles (Fig. 8, reaction *ii*). Finally, addition of DNA-SC-TRAIL conjugates to the final DNA nanostructures in excess facilitates functionalization of the nanostructure (Fig. 8, reaction *iii*).

2. Conjugation was performed by incubating 50-μM SC-TRAIL, 1-mM functional peptides containing FITC (as a positive control) or azide, 25-μM sortase 5 M, in sortase buffer. Martijn Verdoes (Radboud University) kindly gifted the plasmid encoding for Sortase 5 M.
3. The sequence of the functional peptide was NH_2-GGG-K(X)-G-H, where X is the FITC fluorophore or N_3 azide on the lysine sidechain.
4. The reaction mixture was incubated at 4 °C, mixing for total of 2 h, followed by centrifugation using Amicon Ultra-0.5 50 kDa (10 min, 3000 g, 4 °C) to remove the sortase enzyme and the excess peptides.
5. The reaction was monitored with SDS-PAGE Bolt™ 4–12%, Bis-Tris, 1.0 mm from Thermo Fisher Scientific, with a running scheme of 200 V for 35 min.
6. Proteins were stained with SimplyBlue™ SafeStain from Thermo Fisher Scientific (Fig. 9a), while the fluorescent signal from the FITC fluorophore could be measured with excitation and emission at 490 nm and 525 nm, respectively (*see* **Note 4**). Overlap of the fluorescent bands with the TRAIL bands confirmed the successful modification (Fig. 9a).
7. Separately, an azide-containing peptide was conjugated to the SC-TRAILs, which will be used to conjugate a small oligo to allow functionalization of the DNA nanostructure with the ligands.

3.11 Click Reaction to Link SC-TRAIL to DNA

1. 1.5 mg of Dibenzocyclooctyne-N-hydroxysuccinimidyl (DBCO) ester was weighed and dissolved in 300 μL DMSO, and this solution was further used to prepare DBCO-

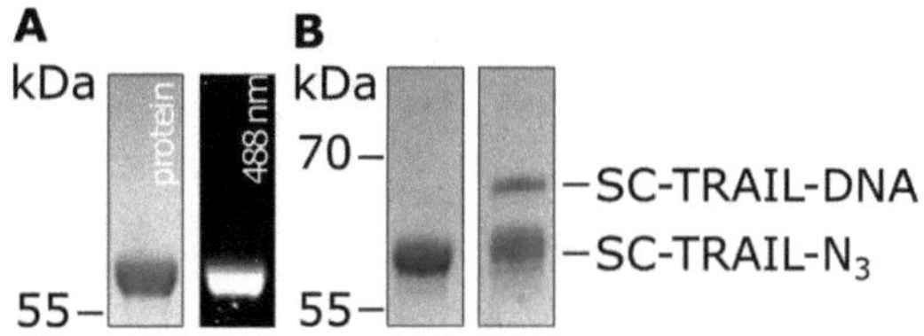

Fig. 9 Conjugation of TRAIL proteins with functional groups. (**a**) SDS PAGE analysis of SC-TRAIL conjugation with a fluorescent peptide, GGG-K (FITC)-G. Overlap of fluorescent bands with SC-TRAIL bands confirms the introduction of the fluorescent peptide. (**b**) SDS PAGE analysis of SC-TRAIL conjugation with the DNA handle. The shift in the band size when the reaction is performed confirms the conjugation of the DNA handle. The band on the left represents the SC-TRAIL-sortag, the band on the right bottom represents SC-TRAIL-N_3, and the band on the right top represents the conjugation product SC-TRAIL-DNA

functionalized DNA handles by mixing 20 μL DBCO ester, 20 μL DMSO, 20 μL PBS, and 20 μL amino-modified DNA handles.

2. The mixture was incubated overnight at 37 °C, 1100 rpm.
3. DBCO-modified DNA handles were then purified with Amicon Ultra-0.5 10 kDa (10 min, 13,000 g, RT), and the concentration was determined with the absorbance at 260 nm.
4. SC-TRAIL-N_3 was incubated with the DBCO-functionalized DNA handle at 4 °C overnight with 3-μM ligand and 30-μM oligo.
5. SC-TRAIL-DNA was then purified from the excess DNA-DBCO using Amicon Ultra-0.5 50 kDa (10 min, 3000 g, 4 °C).
6. The reaction was analyzed with an SDS-PAGE to monitor the reaction (Fig. 9b); the shift of the band confirmed successful labeling of the ligands.
7. The ratio between the SC-TRAIL-N_3 and SC-TRAIL-DNA was estimated using the SDS-PAGE, and this ratio was later used to estimate the concentration of SC-TRAIL-DNA after determination of the concentration with absorbance at 280 nm.

3.12 Functionalization of DNA Nanostructures with SC-TRAIL

1. The DNA-origami nanostructures (50 nM) were incubated overnight at room temperature with 24× concentration of SC-TRAIL-DNA (as there are 24 binding sites).
2. This was followed by purification with Amicon Ultra-0.5100 kDa filters (4 min, 3000 g, RT).
3. The final concentration was determined by absorbance at 260 nm.

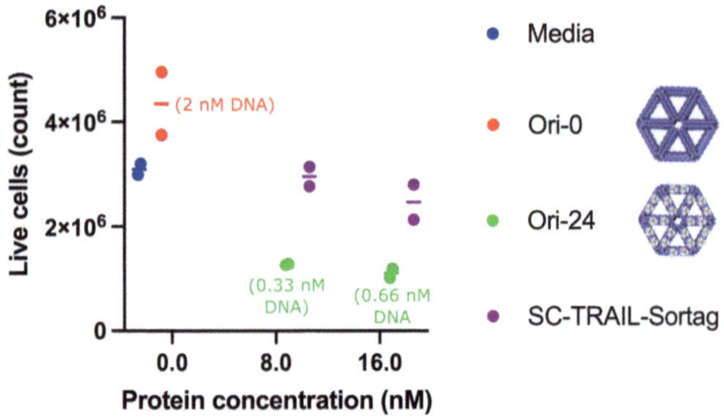

Fig. 10 Super-clustering SC-TRAIL using DNA-nanotechnology results in enhanced cell killing. Live cell counts of Jurkat cells treated with DNA origami constructs containing SC-TRAIL or SC-TRAIL-sortag at different concentrations. DNA concentrations are shown in parentheses

3.13 Cell Killing Induced by DNA-Templated SC-TRAIL

1. Jurkat cells were seeded 100×10^3 cells/well in a 96-well plate in 45 or 20 μL overnight at 37 °C, 5% CO_2.
2. Cells were treated with different concentrations of SC-TRAIL containing DNA origami nanostructures and incubated at 37 °C, 5% CO_2 overnight.
3. Viable cells were counted next day using TC20 Automated Cell Counter from Bio-Rad with trypan blue staining.
4. Figure 10 shows the cell killing properties of the constructs with the SC-TRAIL on TRAIL-sensitive Jurkat cells. For this demonstration, we compared the cell killing of an empty DNA origami construct (red Ori-0; no protein, 2-nM DNA concentration) and a construct with 24 SC-TRAIL binding positions (green Ori-24; 8- and 16-nM SC-TRAIL on 0.33- and 0.66-nM DNA concentration, respectively) at two different concentrations. As Fig. 10 shows, empty DNA origami construct does not induce any cell killing compared to the media control. However, the number of live cells decreases drastically when the construct with Ori-24 is used. Interestingly, SC-TRAIL-sortag alone (purple) does not induce efficient cell killing. This therefore demonstrates the additional efficacy resulting from TRAIL-induced DR5 super-clustering using protein engineering and DNA-mediated nanopatterning.

4 Notes

1. Use UCSF Chimera to open these, do not use UCSF ChimeraX, as this will not display the numbering correctly.
2. There are other important parameters to be aware of, including the length of the staple strand (ideally 20–60 bases long), the proximity of adjacent Holliday junctions (not too close, not

too far), and the directionality (i.e., whether to extend the 3′ or 5′ of the DNA). We do not cover these considerations here, and we instead refer to reference [28] for more information.

3. Different approaches can be used to functionalize the DNA nanostructures, for example, noncovalent binding by DNA hybridization [25], covalent binding by click chemistry [42], using the interaction between biotin and streptavidin [43], or intercalation ability of doxorubicin can be exploited to load drugs [44, 45], also reviewed in [46]. Oftentimes, because of its ease in use and ready availability, the interaction between streptavidin and biotin is used to functionalize DNA nanostructures. However, we avoid using streptavidin for functionalization, because of the fact that streptavidin has four binding domains for biotin and can aggregate DNA nanostructures by binding to multiple biotins from multiple DNA nanostructures.
4. The fluorescent signal has to be measured before the staining with SimplyBlue™ SafeStain to visualize the signal. Oftentimes, the fluorescent signal from the excessive FITC-peptide saturates the fluorescent signal, and to solve this problem, lower part of the gel can be cut to remove the fluorescent peptides.

References

1. Zhang K, Gao H, Deng R, Li J (2019) Emerging applications of nanotechnology for controlling cell-surface receptor clustering. Angew Chem Int Ed 58:4790–4799. https://doi.org/10.1002/anie.201809006
2. Geiger B, Spatz JP, Bershadsky AD (2009) Environmental sensing through focal adhesions. Nat Rev Mol Cell Biol 10:21–33. https://doi.org/10.1038/nrm2593
3. He X-P, Tian H (2018) Lightening up membrane receptors with fluorescent molecular probes and supramolecular materials. Chem 4: 246–268. https://doi.org/10.1016/j.chempr.2017.11.006
4. Jonathan J, Huang T-H et al (2014) Apo2L/TRAIL and the death receptor 5 agonist antibody AMG 655 cooperate to promote receptor clustering and antitumor activity. Cancer Cell 26:177–189. https://doi.org/10.1016/j.ccr.2014.04.028
5. Ashkenazi A (2002) Targeting death and decoy receptors of the tumour-necrosis factor superfamily. Nat Rev Cancer 2:420–430. https://doi.org/10.1038/nrc821
6. Dostert C, Grusdat M, Letellier E, Brenner D (2019) The TNF family of ligands and receptors: communication modules in the immune system and beyond. Physiol Rev 99: 115–160. https://doi.org/10.1152/physrev.00045.2017
7. Pan L, Fu T-M, Zhao W et al (2019) Higher-order clustering of the transmembrane anchor of DR5 drives signaling. Cell 176:1477–1489.e1414. https://doi.org/10.1016/j.cell.2019.02.001
8. Iskratsch T, Wolfenson H, Sheetz MP (2014) Appreciating force and shape — the rise of mechanotransduction in cell biology. Nat Rev Mol Cell Biol 15:825–833. https://doi.org/10.1038/nrm3903
9. Humphrey JD, Dufresne ER, Schwartz MA (2014) Mechanotransduction and extracellular matrix homeostasis. Nat Rev Mol Cell Biol 15: 802–812. https://doi.org/10.1038/nrm3896
10. Wajant H (2015) Principles of antibody-mediated TNF receptor activation. Cell Death Differ 22:1727–1741. https://doi.org/10.1038/cdd.2015.109
11. Kucka K, Wajant H (2021) Receptor oligomerization and its relevance for signaling by receptors of the tumor necrosis factor receptor

superfamily. Front Cell Dev Biol 8:615141. https://doi.org/10.3389/fcell.2020.615141
12. de Bruyn M, Bremer E, Helfrich W (2013) Antibody-based fusion proteins to target death receptors in cancer. Cancer Lett 332: 175–183. https://doi.org/10.1016/j.canlet.2010.11.006
13. Kaplan-Lefko PJ, Graves JD, Zoog SJ, Pan Y, Wall J, Branstetter DG, Moriguchi J, Coxon A, Huard JN, Xu R, Peach ML, Peach ML, Juan G, Kaufman S, Chen Q, Bianchi A, Kordich JJ, Ma M, Foltz IN, Gliniak BC (2010) Conatumumab, a fully human agonist antibody to death receptor 5, induces apoptosis via caspase activation in multiple tumor types. Cancer Biol Ther 9(8):618–631. https://doi.org/10.4161/cbt.9.8.11264
14. Friedl P, Den Boer AT, Gunzer M (2005) Tuning immune responses: diversity and adaptation of the immunological synapse. Nat Rev Immunol 5:532–545. https://doi.org/10.1038/nri1647
15. Wang Y, Baars I, Fördös F, Högberg B (2021) Clustering of death receptor for apoptosis using nanoscale patterns of peptides. ACS Nano 15:9614–9626. https://doi.org/10.1021/acsnano.0c10104
16. Swers JS, Grinberg L, Wang L et al (2013) Multivalent scaffold proteins as superagonists of TRAIL receptor 2–induced apoptosis. Mol Cancer Ther 12:1235–1244. https://doi.org/10.1158/1535-7163.mct-12-1107
17. Schneider H, Yanakieva D, Macarrón A et al (2019) TRAIL-inspired multivalent dextran conjugates efficiently induce apoptosis upon DR5 receptor clustering. Chembiochem 20: 3006–3012. https://doi.org/10.1002/cbic.201900251
18. Lamanna G, Smulski CR, Chekkat N et al (2013) Multimerization of an apoptogenic TRAIL-mimicking peptide by using adamantane-based dendrons. Chem Eur J 19: 1762–1768. https://doi.org/10.1002/chem.201202415
19. Jiang T, Sun W, Zhu Q et al (2015) Furin-mediated sequential delivery of anticancer cytokine and small-molecule drug shuttled by graphene. Adv Mater 27:1021–1028. https://doi.org/10.1002/adma.201404498
20. Batard P, Peterson DA, Devêvre E et al (2006) Dextramers: new generation of fluorescent MHC class I/peptide multimers for visualization of antigen-specific CD8+ T cells. J Immunol Meth 310:136–148. https://doi.org/10.1016/j.jim.2006.01.006
21. Altman JD, Moss PAH, Goulder PJR et al (1996) Phenotypic analysis of antigen-specific T lymphocytes. Science 274:94–96. https://doi.org/10.1126/science.274.5284.94
22. Ranzinger J, Krippner-Heidenreich A, Haraszti T et al (2009) Nanoscale arrangement of apoptotic ligands reveals a demand for a minimal lateral distance for efficient death receptor activation. Nano Lett 9:4240–4245. https://doi.org/10.1021/nl902429b
23. Rothemund PWK (2006) Folding DNA to create nanoscale shapes and patterns. Nature 440:297–302. https://doi.org/10.1038/nature04586
24. Douglas SM, Dietz H, Liedl T et al (2009) Self-assembly of DNA into nanoscale three-dimensional shapes. Nature 459:414–418. https://doi.org/10.1038/nature08016
25. Comberlato A, Koga MM, Nüssing S et al (2022) Spatially controlled activation of toll-like receptor 9 with DNA-based nanomaterials. Nano Lett 22:2506–2513. https://doi.org/10.1021/acs.nanolett.2c00275
26. Stephanopoulos N, Freeman R, North HA et al (2015) Bioactive DNA-peptide nanotubes enhance the differentiation of neural stem cells into neurons. Nano Lett 15:603–609. https://doi.org/10.1021/nl504079q
27. Hellmeier J, Platzer R, Eklund AS et al (2021) DNA origami demonstrate the unique stimulatory power of single pMHCs as T cell antigens. Proc Natl Acad Sci 118:e2016857118. https://doi.org/10.1073/pnas.2016857118
28. Wagenbauer KF, Engelhardt FAS, Stahl E et al (2017) How we make DNA origami. ChemBioChem 18:1873–1885. https://doi.org/10.1002/cbic.201700377
29. Siegemund M, Pollak N, Seifert O et al (2012) Superior antitumoral activity of dimerized targeted single-chain TRAIL fusion proteins under retention of tumor selectivity. Cell Death Dis 3:e295. https://doi.org/10.1038/cddis.2012.29
30. Pettersen EF, Goddard TD, Huang CC et al (2021) UCSF chimeraX: structure visualization for researchers, educators, and developers. Prot Sci 30:70–82. https://doi.org/10.1002/pro.3943
31. Chin SM, Kimberlin CR, Roe-Zurz Z et al (2018) Structure of the 4-1BB/4-1BBL complex and distinct binding and functional properties of utomilumab and urelumab. Nat Commun 9:4679. https://doi.org/10.1038/s41467-018-07136-7
32. Hymowitz SG, Christinger HW, Fuh G et al (1999) Triggering cell death: the crystal structure of Apo2L/TRAIL in a complex with death receptor 5. Mol Cell 4:563–571. https://doi.org/10.1016/S1097-2765(00)80207-5

33. Jun H, Wang X, Parsons Molly F et al (2021) Rapid prototyping of arbitrary 2D and 3D wireframe DNA origami. Nucleic Acids Res 49: 10265–10274. https://doi.org/10.1093/nar/gkab762
34. Jun H, Shepherd TR, Zhang K et al (2019) Automated sequence design of 3D polyhedral wireframe DNA origami with honeycomb edges. ACS Nano 13:2083–2093. https://doi.org/10.1021/acsnano.8b08671
35. Veneziano R, Ratanalert S, Zhang K et al (2016) Designer nanoscale DNA assemblies programmed from the top down. Science 352:1534–1534. https://doi.org/10.1126/science.aaf4388
36. METIS – three-dimensional, algorithmically-generated library of dna origami shapes.
37. TALOS – three-dimensional, algorithmically-generated library of DNA origami shapes.
38. Pettersen EF, Goddard TD, Huang CC et al (2004) UCSF chimera?A visualization system for exploratory research and analysis. J Comp Chem 25:1605–1612. https://doi.org/10.1002/jcc.20084
39. Douglas SM, Marblestone AH, Teerapittayanon S et al (2009) Rapid prototyping of 3D DNA-origami shapes with caDNAno. Nucleic Acids Res 37:5001–5006. https://doi.org/10.1093/nar/gkp436
40. Ma N, Cheng K, Feng Q et al (2023) Nanoscale organization of TRAIL trimers using DNA origami to promote clustering of death receptor and cancer cell apoptosis. Small 19: 2206160. https://doi.org/10.1002/smll.202206160
41. Shi J, Kundrat L, Pishesha N et al (2014) Engineered red blood cells as carriers for systemic delivery of a wide array of functional probes. Proc Natl Acad Sci 111:10131–10136. https://doi.org/10.1073/pnas.1409861111
42. Veneziano R, Moyer TJ, Stone MB et al (2020) Role of nanoscale antigen organization on B-cell activation probed using DNA origami. Nat Nanotech 15:716–723. https://doi.org/10.1038/s41565-020-0719-0
43. Mela I, Vallejo-Ramirez PP, Makarchuk S et al (2020) DNA nanostructures for targeted antimicrobial delivery. Angew Chem Int Ed 59: 12698–12702. https://doi.org/10.1002/anie.202002740
44. Wang Z, Song L, Liu Q et al (2021) A tubular DNA Nanodevice as a siRNA/Chemo-drug co-delivery vehicle for combined cancer therapy. Angew Chem Int Ed 60:2594–2598. https://doi.org/10.1002/anie.202009842
45. Ge Z, Guo L, Wu G et al (2020) DNA origami-enabled engineering of ligand–drug conjugates for targeted drug delivery. Small 16:1904857. https://doi.org/10.1002/smll.201904857
46. Madsen M, Gothelf KV (2019) Chemistries for DNA nanotechnology. Chem Rev 119:6384–6458. https://doi.org/10.1021/acs.chemrev.8b00570

Chapter 5

Controlling the Potency of T Cell Activation Using an Optically Tunable Chimeric Antigen Receptor

Muna Fuyal and John R. James

Abstract

The ability of biological systems to convert inputs from their environment into information to guide future decisions is central to life and a matter of great importance. While we know the components of many of the signaling networks that make these decisions, our understanding of the dynamic flow of information between these parts remains far more limited. T cells are an essential white blood cell type of an adaptive immune response and can discriminate between healthy and infected cells with remarkable sensitivity. This chapter describes the use of a synthetic T-cell receptor (OptoCAR) that is optically tunable within cell conjugates, providing control over the duration, and intensity of intracellular T-cell signaling dynamics. Optical control can also provide control over signaling with high spatial precision, and the OptoCAR is likely to find application more generally when modulating T-cell function with imaging approaches.

Key words Cell signaling, Optogenetics, Antigen receptors, T cells, Transcription factors

1 Introduction

Cells receive, process, and respond to environmental cues that inform and guide their future actions. The processing of these signals primarily happens through intracellular signaling pathways and gene regulatory networks, which are formed from a range of biomolecules. These component parts form pathways that can process, integrate, and transduce information to make crucial cellular decisions, such as differentiation, proliferation, activation, polarization, or apoptosis [1]. Biochemical reactions and molecular interactions form the direct or indirect connectivity in the signaling pathways through network motifs. Identifying these motifs in the complex network of signaling pathways in cells is pivotal in understanding cellular information processing and its ultimate response. The approach of decoding these information networks in cells through gene knockouts or overexpression has provided great insight into cell behavior but does not take into account the

Christoph Wülfing and Robert F. Murphy (eds.), *Imaging Cell Signaling*, Methods in Molecular Biology, vol. 2800,
https://doi.org/10.1007/978-1-0716-3834-7_5,

dynamic nature of cellular processes in time and space. Using light as a signal stimulant provides rapid, targeted, and reversible modulation of gene expression and protein function in cells in a noninvasive manner and with precise spatiotemporal control [2]. A plethora of optogenetic tools have been developed and applied in bacteria [3], yeast [4, 5], mammalian cells [6, 7], and in vivo for a myriad of applications such as control systems in cells for protein function modulation [8], localization [9] and stability [10], organelle motility [11], and reduction of cell-to-cell variability in gene expression [12], light switchable phases in microbial system for optimal metabolite production [13], and drug discovery and translational medicine [14, 15].

Immune cells have similarly intricate signaling networks that work in combination to eliminate disease-causing pathogens and dysfunctional cells from the host. T cells are white blood cells that form part of the adaptive immune system that recognize and respond to processed antigens presented by antigen-presenting cells (APCs) through its antigen receptor (TCR). The TCR is a complex of heterodimeric αβ chains that recognize peptide-MHC complex on APCs and invariant subunits (CD3ϵ, δ, γ, and ζ) responsible for signal transmission to downstream pathways for activation of T cells. Engagement of TCRαβ with peptide-MHC complexes induces the activation of SRC family kinases such as LCK and FYN that can initiate proximal signaling through phosphorylation of immunoreceptor tyrosine-based activation motifs (ITAMs) present in the cytosolic domains of the CD3 proteins. This phosphorylation can subsequently recruit and activate ZAP70, a SYK family kinase, leading to activation of a cascade of accessory proteins. TCR activation signal is further propagated downstream through different signaling pathways such as the Ca^{2+}-NFAT signaling pathway, mitogen-activated protein kinase (MAPK), and nuclear factor-κB (NF-κB) signaling pathways, leading to the activation and expression of transcription factors that are critical for gene expression and functional T-cell response [16]. In this chapter, we will describe a protocol to quantify the activation response of T cells using a light-controllable chimeric antigen receptor (OptoCAR). The OptoCAR construct recapitulates the function of the TCR with intracellular CD3ζ containing ITAMs and extracellular FKBP-FRB based heterodimerization system for precise temporal control over initiation of receptor activation [17]. OptoCAR incorporates the LOVTRAP [18] system for light-based control of the receptor, where illumination with blue light (450 nm) causes the reversible dissociation of an engineered Zdk domain from the LOV2 domain derived from *Avena sativa* (Fig. 1). Zdk is anchored at the plasma membrane using N-terminal myristoylation motif, which diffuses away laterally in the plasma membrane when dissociated from the LOV2 domain and is fused via a short linker to CD3ζ at the C-terminus. In the absence of blue light, T

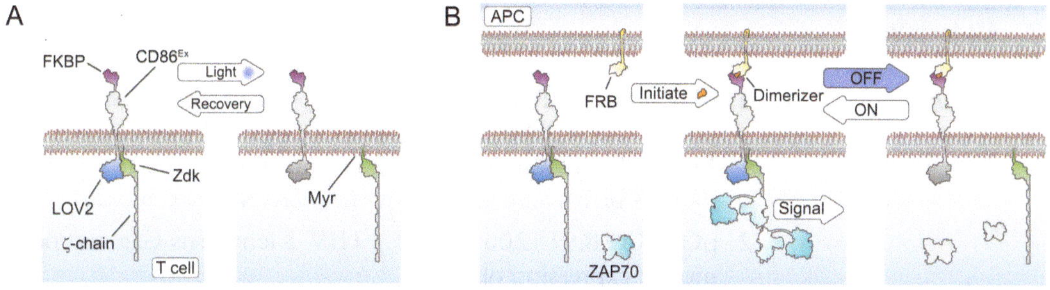

Fig. 1 An optically modulated chimeric antigen receptor (OptoCAR) for T cells (**a**) A schematic of the bipartite OptoCAR complex formed with the component parts labeled. Light-mediated disruption causes dissociation of intracellular signaling tail (CD3ζ-chain) from the extracellular binding domain. (**b**) OptoCAR-T cell engagement with an antigen-presenting cell (APC) does not initiate signaling in the absence of the dimerizer small molecule. OptoCAR signaling from within these cell conjugates is modulated by transitioning from the dark (signaling-competent) state to the "OFF" state by illuminating cells with blue light. (Figure is adapted from Ref. [17])

cell signal transduction from receptor to downstream pathways is initiated by addition of a heterodimerizing drug that links the ligand presented by APC to the ectodomain of the OptoCAR. When the cells are illuminated and the Zdk diffuses away from signal receiving domain, CD3ζ and associated signaling complexes like ZAP70 are inactivated by the presence of highly abundant phosphatases, such as CD45, in the membrane thus inhibiting downstream signal transduction. Using this OptoCAR system, we have previously modulated the intensity, frequency, and duration of T-cell activation to investigate the capability of T cells to integrate signals from multiple simulations and the quantify the persistence of signal information in different nodes of T-cell activation pathways [17].

In order to investigate the optical control over this signal transduction process in T cells, here we will present the protocol to study two functional outputs of receptor activation using J. NFAT cells, which are derived from the parental Jurkat T-cell line [19]. Upon receptor stimulation in J.NFAT cells, the cytoplasmic form of the NFAT transcription factor is dephosphorylated and translocates to the nucleus [20], where it promotes the expression of GFP through binding to response elements upstream of a minimal promoter. The second functional output from the activation of J.NFAT cells is the surface expression of an early activation marker, CD69. To perform the experiment, J.NFAT cells and Raji-FRB^{ex} cells were conjugated in the presence of the heterodimerizing drug in a modified 96-well plate called OptoPlate that allows programmable control of light intensity and duration independently on each well of the 96-well plate.

2 Materials

2.1 Plasmids

1. pHR-OptoCAR-mScarlet (200 ng/mL): This plasmid contains FKBP-CD86(Ex-TM)-LOV2 (ligand binding part) and Myr-Zdk-CD3ζ-mScarlet (signal transmitting part) separated by P2A bicistronic linker sequence. *See* **Notes 1** and **2**.
2. pCMVΔR8.91 (200 ng/mL): HIV-1 lentivirus Gag and polymerase expression plasmid.
3. pMD2.G (100 ng/mL): VSV-G envelope expressing plasmid.

2.2 Cell Culture

1. Cell culture vessels: T75 flasks, T25 flasks, 24-well plates, 6-well plates, round-bottomed 96-well plates, flat-bottomed 96-well plates, μClear flat-bottomed black 96-well plates. *See* **Note 3**.
2. RPMI-1640 media (Gibco) supplemented with 10% heat inactivated fetal bovine serum (HI FBS), 10-mM HEPES buffer and penicillin-streptomycin-glutamine (PSG) cocktail at working concentration (this complete media is called JMEM).
3. Dulbecco's Modified Eagle's Medium, DMEM supplemented with 10% HI FBS and PSG (this complete media is called TMEM).
4. 0.05% Trypsin solution.
5. Sterile phosphate-buffered saline (PBS).
6. HEK293T human embryonic kidney cells.
7. Raji human B lymphocytes.
8. J.NFAT cells: Jurkat T cell lines that express GFP under the control of NFAT promoter following TCR stimulation and constitutively express iRFP713 fluorophore [19].
9. Raji-FRB^{ex} cells: Raji cell lines that constitutively express FRB domain on cell surface and mTagBFP2 fluorophore in cytosol [21].
10. GeneJuice™ (Merck Millipore) or similar transfection reagent.
11. Serum-free medium.

2.3 OptoPlate Instrumentation

To modulate the T-cell response with light stimulation, we used an open-source device called an "OptoPlate" [22], which provides independent illumination of each well of a 96-well plate. For the LOVTRAP system, we used the OptoPlate with two blue (470 nm) LEDs. Detailed instructions for construction of the OptoPlate can be found in this reference [22], and a graphical user interface to facilitate its programming can be found in this reference [23]. *See* **Note 4**.

2.4 Reagents

1. A/C heterodimerizer (previously known as AP21967), from Takara (Cat No: 635057).
2. Human TruStain FcX, which blocks Fc receptor-mediated binding of antibodies.
3. Alexa Fluor 647 anti-CD69 antibody (commercial, we used Biolegend).
4. FACS Wash buffer; 2% (v/v) HI FBS, 0.1% (w/v) NaN_3 in PBS, pH 7.4.
5. FACS Fix; 1.6% (v/v) formaldehyde, 2% (w/v) glucose, 0.1% (w/v) NaN_3 in PBS, pH 7.4.

3 Methods

3.1 Lentiviral Transduction of OptoCAR in Jurkat Cells

1. Plate out 1×10^6 HEK293T cells per well of a tissue-culture treated 6-well plate and allow to adhere overnight at 37 °C and 5% CO_2 in a humidified incubator.
2. Check the confluency of the HEK293T cells in the 6-well plate under the microscope. If the confluency of HEK cells is between 50% and 80%, prepare the following transfection mix:
   - In a 1.5-mL Eppendorf tube (Tube A), add 22-μL sterile water, 6-μL pMD2.G, 6-μL pCMVΔR8.91, and 6-μL pHR-OptoCAR-LOV2-mScarlet for transfecting two wells.
   - In another 1.5-mL Eppendorf tube (Tube B), add 200 μL of serum-free media (SFM) and then 9 μL of Gene Juice transfection reagent.
   - Collect all 40 μL of mix prepared in Tube A using a P200 pipette and transfer it to Tube B.
   - Leave this transfection mix at room temperature for 15 min.

   After this incubation, gently pipette up and down before adding 120 μL of the transfection mix in a drop-wise fashion to two wells of the 6-well plate containing HEK293T cells. Put the plate back in the humidified incubator for 48 h.
3. After 48 h of incubation, collect the supernatant from the two wells, which should now contain lentiviral particles (Caution!) and pipette into 2-mL screw-cap tubes and centrifuge at 800 g for 3 min. This low spin removes debris and cells, and so the pellet should be left undisturbed.
4. Count the number of J.NFAT cells using a hemocytometer (or other means) and transfer 1 mL of the cell culture to each well of a new 6-well plate at 0.5×10^6 cells/mL density.
5. Following centrifugation, pipette the viral supernatant into the 6-well plate containing J.NFAT cells. The transduction can be titrated by adding different volumes of the viral supernatant; as

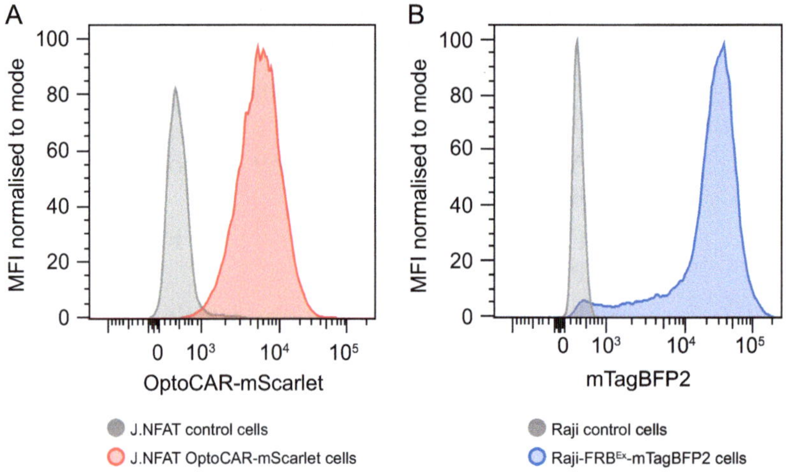

Fig. 2 Expression of OptoCAR-mScarlet in J.NFAT cells and FRB^{ex}-BFP in Raji cells. Following lentiviral transduction of OptoCAR-mScarlet in J.NFAT cells and FRB^{ex}-BFP in Raji cells, cells were tested for the expression of mScarlet in J.NFAT cells (**a**) and mTagBFP2 in Raji cells (**b**) using flow cytometry

an example, use 1000 μL, 500 μL, 250 μL, 100 μL of viral supernatant to different wells of the 6-well plate containing J. NFAT cells and leave a well blank as a control.

6. Add fresh media to each well such that total volume of cell culture is 2 mL and incubate the plate at 37 °C in the humidified incubator overnight.
7. Add 1-mL fresh JMEM medium to each well the following day and incubate the cells.
8. Collect 200 μL of the virally transduced J.NFAT cells in a tube suitable for flow cytometry to check for the transduction efficiency of OptoCAR-mScarlet. Representative flow cytometry data for OptoCAR expression is provided in Fig. 2.

3.2 Setting Up Light-Modulation Experiment

1. Design the light modulation experiment to run on the Opto-Plate. To control the strength of receptor signaling, the intensity of light illuminated on the cells can be varied from 0 μW/cm^2 to 150 μW/cm^2. *See* **Note 5**.
2. Count the number of J.NFAT cells expressing OptoCAR-mScarlet and Raji cells expressing FRB^{ex}-BFP. *See* **Note 6**.
3. Collect 8 × 10^6 J.NFAT-OptoCAR-mScarlet cells in a 50-mL Falcon tube (Tube A) and 14 × 10^6 Raji-FRB^{ex}-BFP cells in another 50-mL Falcon tube (Tube B).
4. Centrifuge both tubes at 800 g for 3 min, discard the supernatant and resuspend the cell pellets in residual media.
5. Add 5.5-mL fresh media to Tube A and mix the cells by pipetting the cell suspension up and down before transferring the cell mix to Tube B.

6. Aliquot the cell mix into ten 1.5-mL Eppendorf tubes and centrifuge at 800 g for 1 min to co-aggregate the cells, and then incubate tube at 37 °C for 10 min. *See* **Note 7**.
7. Resuspend the cell pellets in each tube gently by pipetting up and down and transfer them together to 50-mL Falcon tube.
8. Add 14-mL fresh media to the tube in **Step 7** to make up the volume to 20 mL.
9. To initiate signaling in the cell conjugates, add 2-μM heterodimerizer to the cell mix just prior to aliquoting into the black 96-well plate. For a more detailed protocol, please see Harris et al. [17].
10. Gently mix the cell suspension and transfer to a reagent reservoir. Use a multichannel pipette to transfer 200-μL cell mix to each well in the black μClear 96-well plate (*see* **Note 8**).
11. Place the black 96-well plate on top of the OptoPlate inside the humidified incubator at 37 °C with 5% CO_2, start the program running on the OptoPlate and incubate for the required duration of activation, which is dependent on the program defined on the OptoPlate.

3.3 Processing Samples

Following the activation of J.NFAT-OptoCAR cells as described in Subheading 3.2, collect the cells from the OptoPlate for analysis as described below:

1. Using a multichannel pipette, transfer 200-μL samples from the black 96-well plate to a round-bottom 96-well plate.
2. Centrifuge the 96-well plate at 800 g for 3 min and discard the supernatant by quickly inverting and flicking the plate in a sink.
3. Resuspend the cell pellets in the residual media by pulse vortexing for less than a minute. Be careful to ensure no spillover occurs between wells.
4. Wash the cells by adding 200-μL PBS to the wells in the 96-well plate and centrifuge again at 800 g for 3 min. Discard the supernatant in a similar way as in previous step.
5. Pipette 50 μL of Fc receptor block solution (3-μL Human TruStain FcX in 100-μL PBS) into each well and incubate for 10 min at room temperature.
6. After this incubation, pipette 50 μL of α-CD69 AF647 antibody solution (prepared at 10 μg/mL in FACS Wash buffer) to all wells on the 96-well plate and incubate on ice for 30 min.
7. Add 100-μL FACS Wash buffer to all wells on the 96-well plate and centrifuge at 800 g for 3 min. Discard the supernatant again by quickly inverting and flicking of the plate as before.

8. Wash the cells by adding 200-μL FACS Wash buffer to all wells on the 96-well plate and centrifuge at 800 g for 3 min. Discard the supernatant as before.
9. Fix the cells by adding 200-μL FACS Fix buffer to all wells in the 96-well plate. Leave the plate at 4 °C until analysis.

3.4 Quantifying T-Cell Activation Signal Using Flow Cytometer

1. If available, set up the flow cytometer using a high-throughput (HTS) setup that can accept 96-well plates. If not, the samples in each well must be transferred to tubes compatible with the flow cytometer.
2. For our standard assay, the four fluorescence channels listed below were set up on the flow cytometer:
   - YG561-610/20 (OptoCAR-mScarlet on J.NFAT cells).
   - V405-450/50 (FRB^{ex}-BFP on Raji cells).
   - B488-530/30 (NFAT mediated GFP expression on J. NFAT-OptoCAR).
   - R640-670/30 (CD69-AF647 on J.NFAT-OptoCAR).
3. Set up PMT settings for each fluorescent channel to ensure appropriate dynamic range between positive and negative controls.
4. Run the 96-well plate to collect at least 30,000 gated cells on the flow cytometer.
5. Export the FACS data as .fcs files so the data can be further analyzed.

3.5 Data Analysis

1. Read the .fcs files using suitable software. We used the commercial software FlowJo or MATLAB.
2. Use an appropriate compensation matrix to correct for potential spillover of fluorophores into other channels. *See* **Note 9**.
3. Gate on the live cells using forward scatter (FSC-A) against side scatter (SSC-A) bivariate plots. Depending on what analysis is required, then either gate on the conjugates

   (OptoCAR-mScarlet^{+} Raji-FRB^{ex}-BFP^{+} cells) or singlet J. NFAT cells (OptoCAR-mScarlet^{+} Raji-FRB^{ex}-BFP^{-} cells) using channels YG561-610/20 and V405-450/50, respectively (Fig. 3a).
4. On the gated cells, create histogram plots to show the distribution of expression of GFP using B488-530/30 channel and CD69 using R640-670/30 channel (Fig. 3b). *See* **Note 10**.
5. Export geometric mean (or median) expression of GFP and CD69 and plot against the varying intensity of light (Fig. 4).

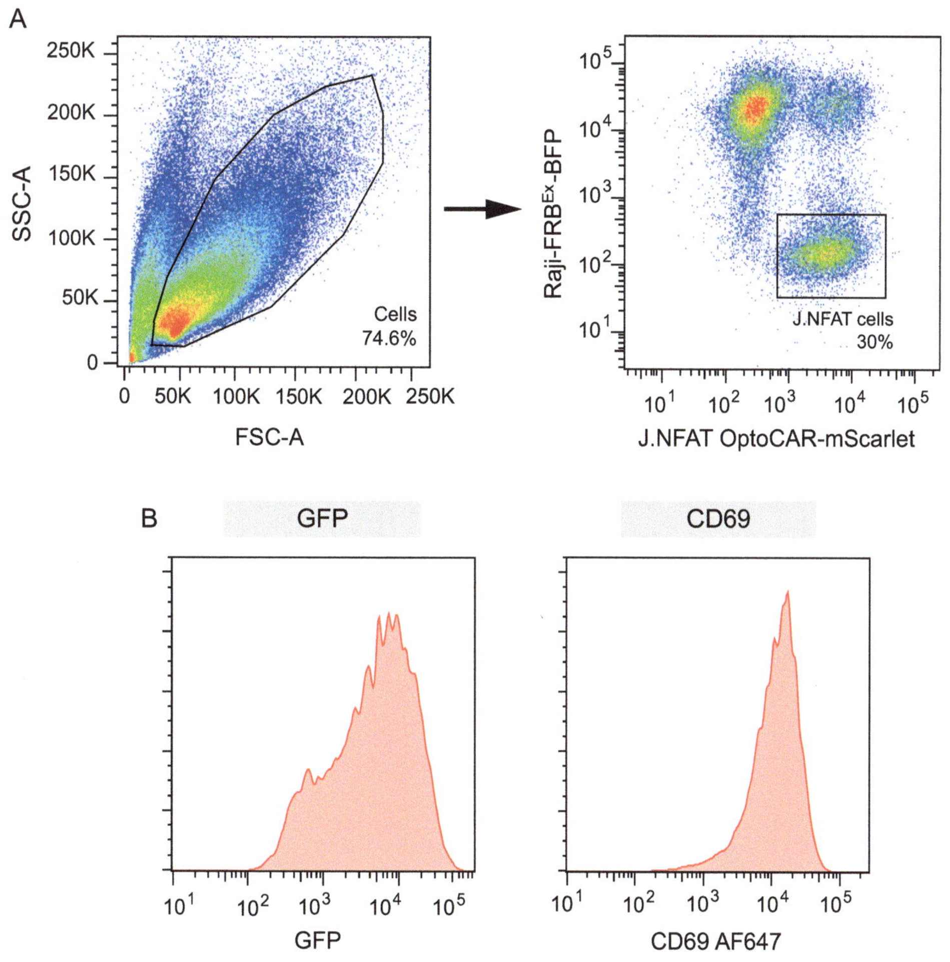

Fig. 3 Analyzing GFP and CD69 expression on activated J.NFAT. (**a**) To analyze the activation response of J.NFAT cells, live cells were first gated on using FSC-A against SSC-A channels. This population was then gated on the J.NFAT cells that express OptoCAR-mScarlet. (**b**) Representative histograms of NFAT-mediated GFP expression or CD69 upregulation on J.NFAT gated cells

4 Notes

1. The construction of this vector has been previously documented [17] and can be obtained from Addgene (#170463).
2. Two light unresponsive forms of pHR-OptoCAR-mScarlet were also constructed by mutations of the LOV2 domain: LOV2^{C450G} remains bound to Zdk even under illumination (always ON), and LOV2^{I539E} has minimal affinity for Zdk even in the dark state (always OFF).
3. Greiner μ-Clear flat-bottomed black 96-well plate fit the dimension of the modified OptoPlate very well in our experience.

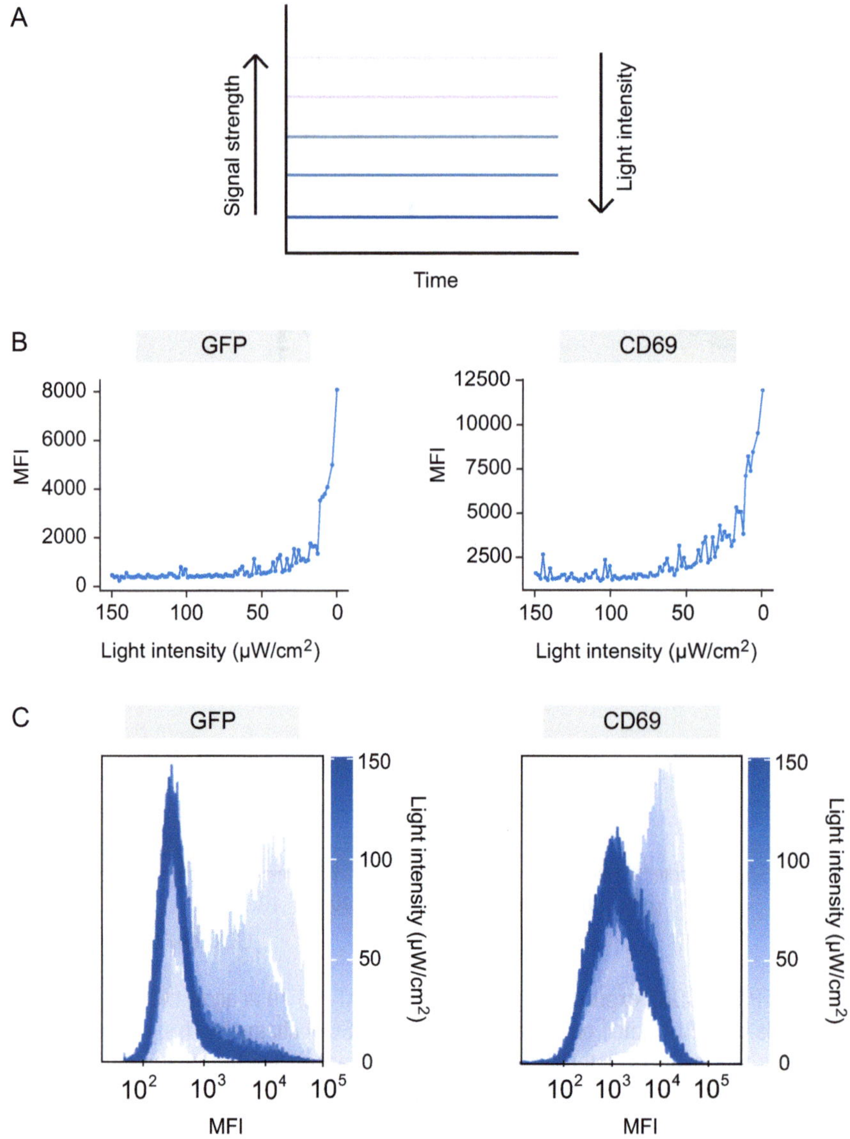

Fig. 4 Distribution of J.NFAT cells following activation by varying intensity of light. (**a**) Signal strength of OptoCAR was titrated by illuminating the conjugated J.NFAT and Raji cells with light intensities varying from 0 to 150 µW/cm^2. (**b**) Line plots show the change in mean fluorescence intensity of NFAT-mediated GFP and CD69 expression with varied intensity of light illuminated on the cells. (**c**) Distribution of cells expressing NFAT-mediated GFP and CD69 when the intensity of light illuminated on the cells were varied

4. The OptoPlate can be modified to improve the uniformity of light distribution across each well by increasing the depth of the plate holder to 18 mm and using 5-mm thick Perspex squares as light diffusers.
5. The maximum light illuminated on the cells should be optimized according to cell lines or primary cells to avoid

phototoxicity. Linear calibration of the arbitrary well intensity values with true LED intensity can be performed with a standard gel-doc imager and has been previously described [17].

6. Count cells at least 2 h before setting up the experiment and keep them at 0.6×10^6 cells/ml density to leave the cells at an optimal state.
7. This brings J.NFAT cells and Raji cells into close apposition before initiating signal transmission by adding heterodimerizer.
8. The media in the wells on the edges of the 96-well plate evaporate more than from the inner wells, so it is better to not use the wells on the edges of the 96-well plate especially if the experiment requires longer incubation.
9. Fluorescence emission from fluorophores such as Alexa Flour 647 can spill over to other channels, depending on the emission filters present on flow cytometer. This needs to be properly compensated with appropriate controls before analyzing the data.
10. Observing changes in the distribution of the activation response when varying cellular stimulation can give more information on cell behavior than simply comparing the mean expression.

References

1. Kholodenko BN (2006) Cell-signalling dynamics in time and space. Nat Rev Mol Cell Biol 7:165–176
2. Mühlhäuser WW, Fischer A, Weber W et al (2017) Optogenetics-bringing light into the darkness of mammalian signal transduction. Biochim Biophys Acta (BBA)-Mol Cell Res 1864:280–292
3. Castillo-Hair SM, Baerman EA, Fujita M et al (2019) Optogenetic control of bacillus subtilis gene expression. Nat Commun 10:1–11
4. Suzuki T, Mioka T, Tanaka K et al (2020) An optogenetic system to control membrane phospholipid asymmetry through flippase activation in budding yeast. Sci Rep 10(1):1–14
5. An-adirekkun J, Stewart CJ, Geller SH et al (2020) A yeast optogenetic toolkit (yotk) for gene expression control in saccharomyces cerevisiae. Biotechnol Bioeng 117:886–893
6. Baaske J, Gonschorek P, Engesser R et al (2018) Dual-controlled optogenetic system for the rapid down-regulation of protein levels in mammalian cells. Sci Rep 8:1–10
7. Yousefi OS, Günther M, Hörner M et al (2019) Optogenetic control shows that kinetic proofreading regulates the activity of the T cell receptor. eLife 8:e42475
8. Wend S, Wagner HJ, Müller K et al (2014) Optogenetic control of protein kinase activity in mammalian cells. ACS Synth Biol 3:280–285. https://doi.org/10.1021/sb400090s
9. Beyer HM, Juillot S, Herbst K et al (2015) Red light-regulated reversible nuclear localization of proteins in mammalian cells and zebrafish. ACS Synth Biol 4:951–958. https://doi.org/10.1021/acssynbio.5b00004
10. Bonger KM, Rakhit R, Payumo AY et al (2014) General method for regulating protein stability with light. ACS Chem Biol 9:111–115
11. Van Bergeijk P, Adrian M, Hoogenraad CC et al (2015) Optogenetic control of organelle transport and positioning. Nature 518:111–114
12. Rullan M, Benzinger D, Schmidt GW et al (2018) An optogenetic platform for real-time, single-cell interrogation of stochastic transcriptional regulation. Mol Cell 70:745–756.e6. https://doi.org/10.1016/j.molcel.2018.04.012
13. Zhao EM, Zhang Y, Mehl J et al (2018) Optogenetic regulation of engineered cellular

metabolism for microbial chemical production. Nature 555:683–687
14. Nguyen NT, Huang K, Zeng H et al (2021) Nano-optogenetic engineering of CAR T cells for precision immunotherapy with enhanced safety. Nat Nanotechnol 16:1424–1434
15. Huang Z, Wu Y, Allen ME et al (2020) Engineering light-controllable CAR T cells for cancer immunotherapy. Sci Adv 6:eaay9209
16. Brownlie RJ, Zamoyska R (2013) T cell receptor signalling networks: branched, diversified and bounded. Nat Rev Immunol 13:257–269
17. Harris MJ, Fuyal M, James JR (2021) Quantifying persistence in the T-cell signaling network using an optically controllable antigen receptor. Mol Syst Biol 17:e10091
18. Wang H, Vilela M, Winkler A et al (2016) LOVTRAP: an optogenetic system for photoinduced protein dissociation. Nat Methods 13:755–758
19. James JR (2018) Tuning ITAM multiplicity on T cell receptors can control potency and selectivity to ligand density. Sci Signal 11:eaan1088
20. Mognol GP, González-Avalos E, Ghosh S et al (2019) Targeting the NFAT: AP-1 transcriptional complex on DNA with a small-molecule inhibitor. Proc Natl Acad Sci 116:9959–9968
21. James JR, Vale RD (2012) Biophysical mechanism of T-cell receptor triggering in a reconstituted system. Nature 487:64–69
22. Bugaj LJ, Lim WA (2019) High-throughput multicolor optogenetics in microwell plates. Nat Protoc 14:2205–2228
23. Thomas OS, Hörner M, Weber W (2020) A graphical user interface to design high-throughput optogenetic experiments with the optoplate-96. Nat Protoc 15:2785–2787

Chapter 6

Imaging Cell Signaling in Tissues Using the IBEX Method

Yuxuan Meng, Changtai Li, Gregory S. Patient, Yuan Tian, and Colin J. Chu

Abstract

The study of cell signaling within tissues can be enhanced using highly multiplexed immunohistochemistry to localize the presence and spatial distribution of numerous pathways of interest simultaneously. Additional data can also be gained by placing the identified proteins into the context of adjacent structures, stroma, and interacting partners. Here, we outline a protocol for using the recently described IBEX method on tissues. This is an open and simple cyclic immunohistochemistry approach suited to this application. We describe a simplified protocol and provide guidance on the method, using a 12-marker panel on human retina to demonstrate the approach.

Key words Cell signaling, IBEX, Immunohistochemistry, Retina, Immunofluorescence

1 Introduction

Classical cell signaling imparts the ability to cells to detect and respond to internal and external environmental stimuli, modulating biological processes to effect a change in function and behavior [1, 2]. Dysregulation of cell signaling pathways is also frequently implicated in various pathological states including autoimmune disorders and cancer [3]. Therefore, studying cell signaling is vital for understanding the molecular mechanisms that control cellular responses and identifying potential therapeutic targets. A huge amount of information can be gained by the in vitro study of cell signaling; however, examining responses at an ex vivo tissue level stands to supplement the integration of signaling across a wider and systems level of understanding. Furthermore, the use of emerging spatial biology techniques will provide additional insights through the ability to localize signaling across tissues.

The Iterative Bleaching Extends Multiplexity (IBEX) method is a recently described imaging approach that can be applied to enable highly multiplexed analysis of signaling proteins in tissue sections

Christoph Wülfing and Robert F. Murphy (eds.), *Imaging Cell Signaling*, Methods in Molecular Biology, vol. 2800, https://doi.org/10.1007/978-1-0716-3834-7_6,

[4, 5]. Multiplexed immunohistochemistry on tissue cryosections can be performed with existing, validated, and commercially available signaling pathway antibodies using IBEX. Only standard fluorescence confocal microscopes are required improving the accessibility to most research groups. IBEX works through registration of iterative cycles of fluorescent antibody staining with bleaching to achieve fluorophore inactivation between each round.

In this chapter, we demonstrate the technique on human retina using selected proteins in the p16, NF-κB, and cGAS-STING [6] signaling pathways that have been studied in our lab in relation to inflammation and neuronal senescence. In addition, we can combine other structural and cellular antibody markers to place the cells with detected signaling proteins in spatial context of the surrounding tissue environment (Fig. 1). IBEX provides a powerful tool to study cell signaling by allowing for the simultaneous detection of multiple targets simultaneously and in tissue to provide a more comprehensive and spatially resolved view of cell signaling networks.

2 Materials

2.1 Biological Materials

Ensure the tissue sourced is covered by appropriate research ethics clearance. Human retina for this protocol was obtained through Moorfields Eye Hospital Biobank (NHS REC approval: 20/SW/0031), with consent and eye retrieval provided by NHS blood and transplant tissue services. The retina used in this study was obtained from a 50-year-old male donor who died of decompensated alcoholic cirrhosis.

2.2 Solutions and Key Reagents

1. Sucrose: Prepare 15% and 30% sucrose solution (w/v) dissolved in PBS. Keep the solution sterile by passing through a 0.22-um filter, and it can be stored at 4 °C for several months. Dilute 30% sucrose in PBS to get 15% sucrose.
2. Fixative: 4% Paraformaldehyde prepared from 16% methanol-free formaldehyde solution in ampules (e.g., Thermo Fisher Scientific, Cat. No. 28908).
3. Post-fixative: 1:4 dilution of BD Cytofix/Cytoperm (BD Biosciences, Cat. No. 554722) in PBS.
4. Embedding solution: Optimal cutting temperature (OCT) medium (CellPath, SKU: KMA-0100-00A).
5. Microscope slide-coating solution: Chrome alum gelatin (Newcomer Supply, Cat. No. 1033B).
6. Blocking solution: 1:20 Normal Donkey Serum 1:100 Human Fc Block (or species equivalent) diluted in 1× BD Perm Wash (BD Biosciences, Cat. No. 554722, made from the 10× stock with PBS).

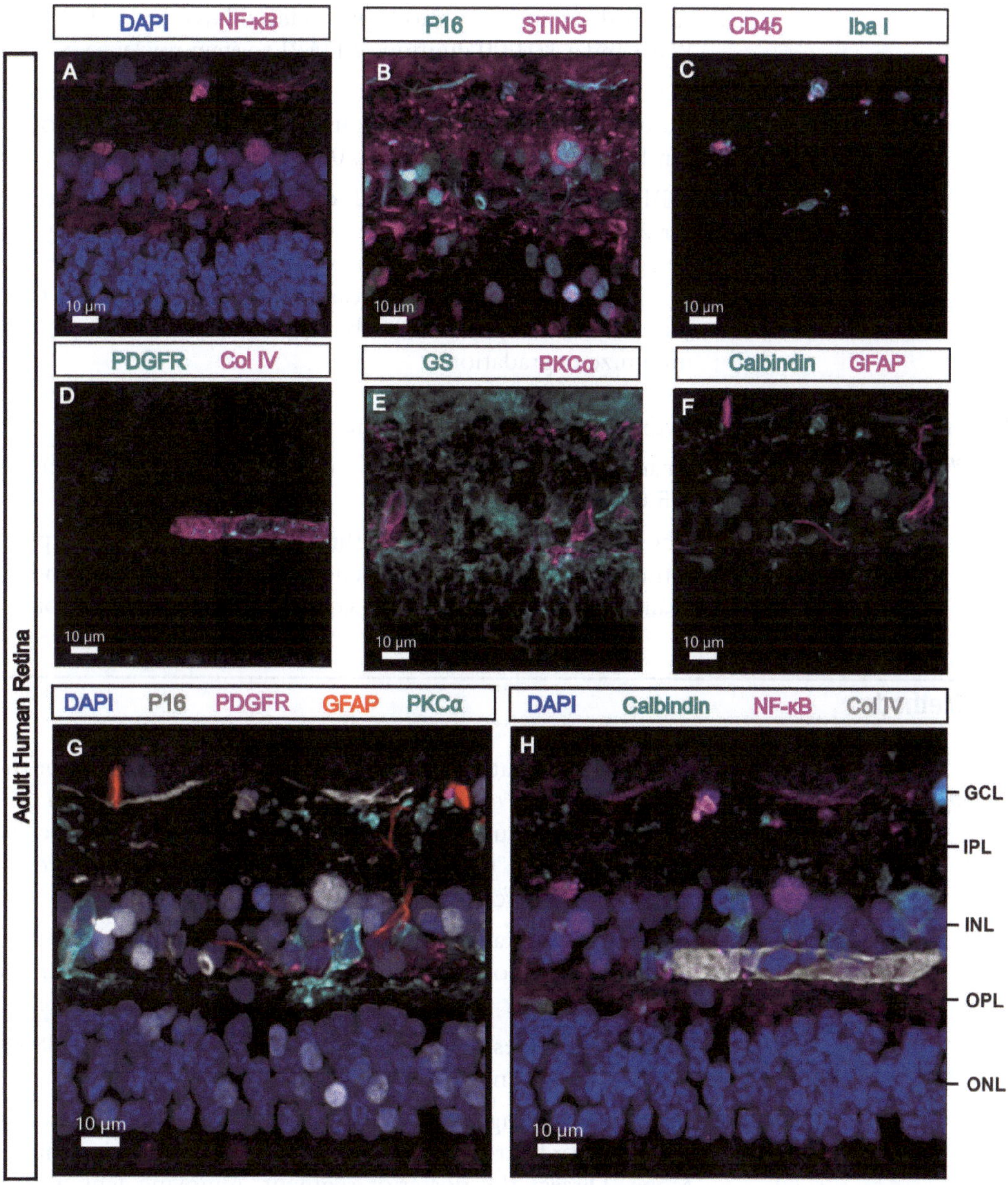

Fig. 1 An exemplar 12-marker IBEX dataset performed on adult human retina. The same location of the sample is shown in all panels. For clarity, two markers per panel are shown in (**a**–**f**). Merged panels of five markers illustrate the combined panel from multiple rounds in (**g**–**h**). A neuronal cell where three different signaling markers are detectable is seen. Other markers identify surrounding retinal structures and cell types that may be interacting. *GCL* ganglion cell layer, *IPL* inner plexiform layer, *INL* inner nuclear layer, *OPL* outer plexiform layer, *ONL* outer nuclear layer

7. Antibody-staining solution: antibodies diluted in 1× BD Perm Wash and a 1:1000 dilution of DAPI to stain nuclei to be used as a fiducial marker.
8. Mounting solution: Fluoromount G mounting media (Southern Biotechnologies, Cat. No. 00-4958-02).
9. $LiBH_4$ solution: 5 mg in 5 mL of distilled water. Leave to rest for 20 min until small bubbles are seen forming. Note hydrogen gas can be generated in an exothermic reaction, so perform this in a chemical hood with care. $LiBH_4$ (e.g., STREM Chemicals, Cat. No. 93-0397) is best purchased in 1-g aliquots to minimize degradation.

2.3 Data Analysis Software

1. Python (version 3.7.0 or higher).
2. Imaris and Imaris Converter (Oxford instruments, version 9.5.0 or higher).
3. The free registration algorithm is available from https://github.com/niaid/sitk-ibex, and the Imaris extension code is available from https://github.com/niaid/imaris_extensions

3 Methods

3.1 Tissue Grossing and Processing

1. Use standard fixation conditions and tissue preparation for the tissue and organ of interest. Here, we freshly prepare a 4% PFA solution and incubate posterior segments supplied by the eye bank in this at 4 °C for 48 h. This is performed in an appropriate biological safety cabinet.
2. Exchange the fixative on the posterior segments for a solution of a 1:4 dilution of BD Cytofix/Cytoperm in PBS and incubate at 4 °C for 16 h.
3. Wash the samples twice at room temperature using 1xPBS for 5 min each to remove excess fixative.
4. Prepare dissecting tools and place the sample on a dissecting mat or other fiber-free clean surface. We use a dissection microscope. Dissect the posterior segment, removing lens, sclera, and vitreous to isolate the retina. Dissect this into pieces around 10 mm by 10 mm.
5. Place the samples in completely dissolved 15% sucrose solution for a few hours at 4 °C until the sample has sunk to the bottom of the container.
6. Transfer the sample to a completely dissolved 30% sucrose solution at 4 °C overnight.
7. Fill a Histomould with optimal cutting temperature (OCT) medium and embed the tissue. Use forceps to adjust the tissue orientation for subsequent frozen sectioning. Ensure the tissue

is located approximately at the center of the mold. Place the Histomould on dry ice and allow to slowly freeze.

8. Label the mold and store in a −80 °C freezer until required.

3.2 Tissue Preparation

1. Coat microscope slides with chrome alum gelatin by adding 5 μL to one side and use a second microscope slide to smear the gelatin evenly along the entire slide in one smooth motion. This is akin to a blood smear. Then dry the coated slide for 1 h in a 60 °C oven.
2. Section the OCT-embedded tissue with a cryostat at 12- to 15-μm thickness on to the center of the coated slides.
3. Outline the tissue with ImmEdge hydrophobic pen and leave to dry overnight at room temperature.

3.3 Permeabilization, Blocking, Staining, and Mounting

1. Wash tissue with PBS for 5 min to remove OCT medium and aspirate this away.
2. Cover the tissue with blocking solution and leave for 1–2 h at room temperature in a humidified chamber.
3. Prepare the panel of antibodies diluted in antibody-staining solution.
4. Aspirate the blocking solution and then apply the first designed primary antibody panel (*see* **Notes 1–6** and Table 1 for details). Incubate overnight at 4 °C in a humidified chamber.
5. Wash with 1xPBS three times then mount the sample with a drop of Fluoromount G mounting media and a coverslip.
6. Allow 1 h for the slide to settle at room temperature before imaging. Keep the time between mounting and imaging consistent between IBEX rounds.

3.4 Imaging with a Confocal Microscope

1. Use a confocal microscope according to standard operating instructions (we employ a Leica TCS SP8 X inverted confocal microscope) to image the fluorophores in the panel designed.
2. Place the slide firmly into the top left corner of the stage insert, so the position remains consistent.
3. Record all the parameters, as these will need to be reloaded at the next imaging session, such as objective, zoom, resolution, z-stack size (number of steps and spacing). Set the z-stack clearly within visible tissue where the fiducial marker (e.g., DAPI) is imaged and distinctive.
4. Acquire and save the image.

3.5 Sample Bleaching

1. Remove the slide and place in a 50-mL falcon tube filled with 1xPBS and wait around 20 min until the coverslip slides off spontaneously.

Table 1
List of antibodies used in the study

Target	Host	Fluorophore	Manufacturer	Catalog number	RRID	Dilution
$p16^{INK4a}$	Rabbit	AF546	Santa Cruz	sc-1661	AB_628067	1:50
STING (D2P2F)	Rabbit	AF647	Cell Signaling technology	13647	AB_2732796	Conjugation kit then 1:25
NF-κB p65	Rabbit	AF546	Santa Cruz	sc-8008	AB_628017	1:25
Iba1	Goat	Purified	Abcam	ab5076	AB_2224402	1:50
Goat IgG (H + L)	Donkey	Secondary AF546	Thermo fisher	A-11056	AB_2534103	1:1000
PDGFR	Rabbit	AF488	Abcam	ab196376	AB_2936929	1:100
GFAP	Rabbit	AF647	Biolegend	837511	AB_2734610	1:200
PKCα	Rabbit	AF680	Santa Cruz	sc-8393	AB_628142	1:100
Glutathione Synthetase	Rabbit	AF488	Proteintech	CL488-66323	AB_2883309	1:100
Calbindin D-28 K	Rabbit	AF550	Millipore Merck	ABN2192	AB_2935805	Conjugation kit then 1:25
CD45	Rabbit	iFluor 594	Caprico biotechnologies	1016136	AB_2936930	1:25
Collagen IV	Rabbit	AF647	Abcam	ab19808	AB_445160	1:50

AF Alexa Fluor™

2. Wash three times for 5 min in 1xPBS.
3. Apply freshly prepared LiBH4 solution to the tissue at room temperature for 15 min exposed to standard room lights. Remove and replace with fresh LiBH4 solution for a further 15 min.
4. Wash the sample with 1xPBS for 5 min three times.
5. Prepare the next panel of antibodies diluted in antibody-staining solution.
6. Aspirate the PBS and then and apply the next antibody panel (*see* **Notes 7–11** and Table 1 for details). Incubate overnight at 4 °C in a humidified chamber.
7. Wash with 1xPBS three times then mount the sample with a drop of Fluoromount G mounting media and a coverslip.
8. Allow 1 h for the slide to settle at RT before imaging. Keep the time between mounting and imaging consistent between IBEX rounds.

3.6 Realignment on Confocal Microscope

1. Replace the slide into the top left corner of the slide insert.
2. Reload the previous parameters to ensure consistent with the last acquisition.
3. Open the image file from Panel 1 and use the fiducial marker (e.g., DAPI) to realign the start and end of the z-stack of the current sample to this using the distinctive nuclear structures manually. This needs to be close, but the registration software will refine the fit in post-processing.
4. Acquire and save the image.
5. Repeat from Subheading 3.5, **step 1** multiple times, until all the desired panels have been imaged.

3.7 Post-processing and Registration

1. Transfer all the confocal imaging files for the multiple rounds to an image processing workstation. Imaris software is recommended for this workflow.
2. Perform initial channel labeling and any pre-processing such as Gaussian smoothing, brightness, and contrast correction required.
3. Label the channels as required by the SITK-IBEX package (e.g., P1_GFAP_AF647) and save the files.
4. Launch and run the SITK-IBEX Python package to align the rounds as per the included standard instructions.
5. The final combined file can then be reviewed in Imaris and taken forward for image analysis as dictated by the scientific aims.

4 Notes

1. Careful planning is required to design panels of markers (antibodies or lectins) to be used with the IBEX technique. The goal is to minimize the overlap of signal between one round to the next and optimize the fit of available fluorophore conjugated antibodies into the minimum number of rounds. The microscope or imager setup will determine how many fluorophores can be detected in a single panel.
2. Ensure planned markers are compatible with IBEX and wherever possible use antibodies directly conjugated to fluorophores.
3. Unconjugated primary antibodies that require a fluorescent secondary antibody can be used but they must be used in rounds in which there are no other antibodies that were raised in the same species and where there have been none in the preceding rounds. This is because IBEX only inactivates

fluorophores, it does not strip antibodies unlike other approaches. Alternatively, purified antibodies can be labeled with commercial antibody conjugation kits.

4. It is advisable to avoid using antibodies that target the same cell type, or tissue region in the same round as this makes it harder to distinguish if there is spectral overlap of fluorophores.
5. Ensure you use antibodies conjugated to fluorophores, which have been validated to bleach in the original protocol [5] or that you have tested the bleaching on.
6. Include a common marker which does not bleach as a fiducial marker throughout all the rounds for the SITK-IBEX registration software. Usually, a reliable nuclear marker such as DAPI is ideal, or use an antibody conjugated to Alexa Fluor 594.
7. Use the smallest number of rounds possible by fitting as many antibodies and different fluorophores into each panel as the microscope setup allows.
8. Place the most abundant and therefore brightest markers, for instance, lectins or highly efficient antibodies in later rounds to minimize leftover signal that can arise from less efficient bleaching due to shear abundance.
9. Place the weakest markers or challenging antibodies (e.g., transcription factor or phosphoprotein directed antibodies) in early panels to increase the likelihood of a detectable signal.
10. Before starting the full IBEX protocol, test each planned panel of antibodies on separate tissue slides to confirm they work individually, titrate the concentrations, and ensure the markers are distinguishable in each round.
11. For additional information and help see the IBEX Knowledge Base and Forum at https://doi.org/10.5281/zenodo.7693278

References

1. Booth LA, Tavallai S, Hamed HA et al (2014) The role of cell signalling in the crosstalk between autophagy and apoptosis. Cell Signal 26(3):549–555
2. Borst J, Ahrends T, Bąbała N et al (2018) CD4+ T cell help in cancer immunology and immunotherapy. Nat Rev Immunol 18(10):635–647
3. Kuchitsu Y, Mukai K, Uematsu R et al (2023) STING signalling is terminated through ESCRT-dependent microautophagy of vesicles originating from recycling endosomes. Nat Cell Biol 25(3):453–466
4. Radtke AJ, Chu CJ, Yaniv Z et al (2022) IBEX: an iterative immunolabeling and chemical bleaching method for high-content imaging of diverse tissues. Nature Protoc 17(2):378–401
5. Radtke AJ, Kandov E, Lowekamp B et al (2020) IBEX: a versatile multiplex optical imaging approach for deep phenotyping and spatial analysis of cells in complex tissues. PNAS 117(52):33455–33465
6. Valls PO, Esposito A (2022) Signalling dynamics, cell decisions, and homeostatic control in health and disease. Curr Opin Cell Biol 75:102066

Chapter 7

Investigating Epidermal Growth Factor Receptor Trafficking with High-Resolution Enzymatic Protein-Tagging and Transmission Electron Microscopy

Nicholas Ariotti

Abstract

Enzymatic ascorbate peroxidase (APEX) tagging allows for high-resolution, three-dimensional protein distribution analyses in cells and tissues. This chapter describes the application of APEX-tagging to visualize the trafficking of the epidermal growth factor receptor (EGFR) during epidermal growth factor-mediated receptor activation. Here, we describe the preparation of cells, methods to validate the stimulation of the EGFR, and visualization of the APEX-resolved distribution of the EGFR in the transmission electron microscope.

Key words Enzymatic-tagging, Electron microscopy, EGF receptor, Receptor trafficking

1 Introduction

The localization of a protein provides critical insight into its role in the cell. Cell biologists localize proteins using microscopy to better understand their function. In basic biodiscovery research, protein localization can inform enrichment with an organelle or provide contextual information regarding interaction partners under different experimental stimuli. In developmental biology, studying protein localization of targets at high-resolution over time can provide crucial insight into the role of a protein during development. Protein localization is also essential for studying how changes in the subcellular distribution of proteins can underpin a transition toward disease states [1, 2]. The higher the resolution of protein localization, the more confident one can be regarding the functional insights in protein behavior.

Traditionally, cell biological protein localization has been performed using fluorescence microscopy (FlM). This method relies upon colocalization of a protein of interest with well-characterized markers of subcellular organelles or with interacting proteins of

Christoph Wülfing and Robert F. Murphy (eds.), *Imaging Cell Signaling*, Methods in Molecular Biology, vol. 2800,
https://doi.org/10.1007/978-1-0716-3834-7_7,

known biological function [3]. While this modality can provide highly sensitivity and dynamic information regarding protein behavior in complex biological systems and in live cells, FlM is still limited by the resolution it can routinely attain [3]. While electron microscopy (EM) does not provide dynamic information (fixation and processing of cells is required to allow the sample to withstand the rigors of the electron beam), it does generate higher-resolution information [4]. Over the last decade, enzymatic ascorbate peroxidase (APEX)-tagging has been developed and applied extensively to perform three-dimensional, highly resolved protein localization for EM [5].

APEX2 is a largely monomeric, low-molecular-weight enzymatic protein tag (28 kDa) derived from soybean that can be conjugated to proteins of interest for localization in cells and tissues [5, 6]. When APEX is expressed and reacted in the presence of 3′ 3′-diaminobenzidine (DAB) and H_2O_2 as a catalyzing substrate, an osmiophilic polymer is formed at the site of the protein tag. In standard biological EM sample preparation, osmium tetroxide (OsO_4) is used as a lipid fixative and contrasting agent. However, when post-fixation with OsO_4 is performed after the DAB reaction, there is a preferential deposition of heavy osmium atoms at the site of the DAB reacted polymer. This preferential deposition increases the scattering potential of the electron beam at the site of the APEX-tag and DAB reaction product to facilitate the visualization of the protein of interest by changes in sample contrast. Direct enzymatic-tagging has been used extensively to characterize the protein distribution in various cell and animal models [6–8]. While this method has been a highly successful strategy for analyzing 3D protein distribution at subcellular resolution, we have previously observed that direct tagging can be disruptive to normal subcellular distribution in certain proteins [9]. As such, we characterized the localization of proteins using a modular approach through the conjugation of the APEX-tag to a nanobody directed against common fluorescence markers attached to the protein of interest (Fig. 1a, b): mCherry and GFP [9, 10]. This work demonstrated that we could detect the expression of proteins of interest rapidly and at low-resolution with FlM and localize proteins at high-resolution in the electron microscope with minimal effects on protein behavior. The utility of modular APEX-tagging has been applied extensively in cells and animal models [9, 10, 11, 12].

The epidermal growth factor receptor (EGFR) is a well-studied receptor tyrosine kinase that functions to translate extracellular signals from ligands into intracellular effector signaling cascades. EGFR is an essential protein involved in the control of cell proliferation and cell death [13]; loss of the receptor is embryonically lethal [14]. Overexpression and mutation of the EGFR is associated with tumorigenesis and metastasis; these mutations can affect the intracellular trafficking of the receptor [13].

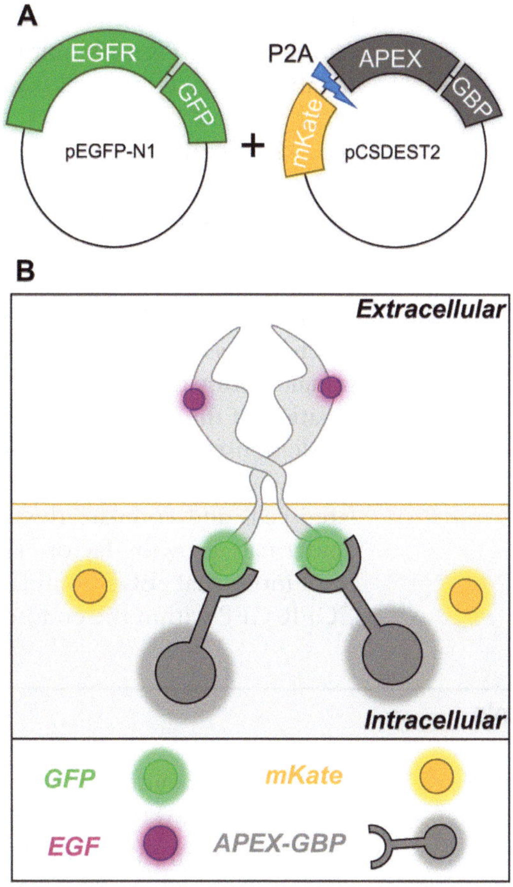

Fig. 1 Schematic of APEX-GBP modular detection of EGFR-GFP. (**a**) Plasmid constructs that are doubly transfected for subcellular EGFR distribution analyses. (**b**) EGFR-GFP dimer activated by ligand binding (AF647-EGF) bound to APEX-GBP. EGFR-GFP fluorescence signal demonstrates expression of EGFR plasmid; mKate cytosolic fluorescence highlights the expression of the APEX-GBP construct. The DAB reaction results in the formation of the brownish polymer that is further contrasted into a black deposit by OsO_4

The EGFR is regulated by a series of conformational changes and oligomerization steps that control the spatial distribution of the receptor at the PM. Ligand binding causes EGFR dimerization and transautophosphorylation and stimulates the recruitment of effector proteins and activation of multiple signaling cascades (Akt, ERK, and JAK/STAT) to effect transcriptional responses [15]. Ligand binding activates the EGFR, and this signaling response is controlled through homeostasis in intracellular trafficking [13, 15].

Activation of EGFR results in rapid endocytosis through clathrin-dependent or clathrin-independent routes [13, 16, 17], whereby EGFR is then sorted toward different intracellular fates to control longer-term signaling responses [18]. For signal termination, the EGFR is ubiquitinated and targeted for lysosomal degradation. For continued signaling, the receptor is sorted for recycling and directed back to the plasma membrane (PM) via recycling endosomes for subsequent rounds of activation [17]. This sorting fate is determined by a complex array of factors including receptor-ligand binding affinity, the co-association of adaptor cargo proteins on the cytoplasmic leaflet of the membrane, and the ubiquitination level of the receptor [13].

Despite intensive studies over decades, the precise molecular mechanisms governing EGFR internalization and trafficking remain poorly defined. Similarly, little is known regarding the organization of the receptor during endosomal sorting. In this chapter, we demonstrate the application of modular APEX-tagging to characterizing the intracellular trafficking of EGFR in control unstimulated cells and compare this distribution to cells stimulated with epidermal growth factor (EGF). We demonstrate at high-resolution that EGF-stimulation results in the enrichment of EGFR-GFP within the endosomal pathway.

2 Materials

2.1 Solutions

1. Phosphate-buffered saline: 137-mM NaCl; 2.7-mM KCl, 10-mM Na_2HPO_4, 1.8-mM KH_2HPO_4, pH 7.4.
2. DMEM cell culture medium: DMEM, 10% fetal bovine serum (FBS), 2-mM L-glutamine (Gibco).

2.2 Plasmids

1. EGFR-EGFP (pEGFP-N1 vector: Addgene plasmid # 32751). The GFP-tag is localized to the intracellular c-terminal kinase domain of the EGFR (Fig. 1a, b). EGFR-GFP was a gift from Alexander Sorkin (Addgene plasmid # 32751; http://n2t.net/addgene:32751; RRID:Addgene_32751) [19].
2. mKate2-P2A-APEX2-GBP (pCSDEST2 vector: Addgene plasmid # 67650). This construct consists of a cytoplasmically localized fluorescent reporter, mKate, that is separated from the APEX-GBP via a P2A self-cleavage domain. This facilitates a direct 1:1 expression ratio of mKate reporter to APEX-GBP. In APEX-GBP the APEX-tag (28 kDa) is directly conjugated to a GFP-nanobody construct (15 kDa) that binds GFP and YFP with high affinity (Fig. 1a, b).

2.3 Fluorescence Microscopy

1. Zeiss LSM 900 Upright Confocal with Airyscan 2 (Carl Zeiss Microscopy GmbH).
2. Zen Blue v3.6 (Carl Zeiss Microscopy GmbH).

3. Alexa Fluor™ 647 Epidermal Growth Factor complex (Thermo Fisher Scientific; E35351).
4. 4% PFA phosphate-buffered saline (pH 7.4): diluted from 16% aqueous paraformaldehyde (PFA) stock solution (Electron Microscopy Sciences: 15700) into PBS.
5. ProLong Glass Antifade Mountant (Thermo Fisher Scientific; 36982).

2.4 Electron Microscopy

1. PELCO Biowave with ColdSpot Pro Chiller system (Ted Pella Inc).
2. Leica EM UC6 Ultramicrotome (Leica Microsystems).
3. DiATOME 45° diamond knife (ProSciTech; UU45-030).
4. 0.2 M sodium cacodylate buffer: Sodium cacodylate trihydrate (Sigma-Aldrich; C0250) dissolved in PBS, pH 7.4.
5. 0.1 M sodium cacodylate buffer: Sodium cacodylate trihydrate (Sigma-Aldrich; C0250) dissolved in PBS, pH 7.4.
6. 2.5% Glutaraldehyde phosphate-buffered saline: diluted from 25% Glutaraldehyde stock solution (Electron Microscopy Sciences; 16220) into 0.1 M phosphate-buffered saline, pH 7.4.
7. 2.5% Glutaraldehyde in sodium cacodylate buffer: diluted from 25% Glutaraldehyde stock solution (Electron Microscopy Sciences; 16220) into 0.1 M sodium cacodylate buffer, pH 7.4.
8. DAB in 0.1 M sodium cacodylate buffer. Dissolve a single 10-mg tablet of 3,3′-diaminobenzidine tetrahydrochloride (DAB) (Sigma-Aldrich; D4168) in water to a concentration of 2 mg/mL. Vortex vigorously until dissolved. Add 0.2-M sodium cacodylate buffer to a final concentration of 1 mg/mL of DAB in 0.1 M sodium cacodylate buffer.
9. 1% OsO_4 in 0.1 M sodium cacodylate buffer: Diluted from 4% aqueous osmium tetroxide (OsO_4) (ProSciTech; C010) in 0.1 M sodium cacodylate buffer.
10. Ethanol series for dehydration in decreasing double distilled water (30%, 50%, 70%, 90%, 100%).
11. LX112 resin (Ladd Research; 21310) infiltration series in decreasing ethanol (33%, 66%, 100%).
12. BEEM capsules (ProSciTech: RB001).

2.5 Electron Microscopy Visualization Equipment and Software

1. JEM-1011 Transmission Electron Microscope (JEOL Japan).
2. Morada CCD camera (EMSiS GmbH).
3. iTEM (EMSiS GmbH).

3 Methods

3.1 Cell Culture

1. Thaw BHK cells into DMEM cell culture medium in a T75 flask and incubate cells overnight in a cell incubator at 37 °C with 5% CO_2.
2. Grow cells to 80–90% confluency, wash cells with PBS, and trypsinize with 0.5% Trypsin-EDTA for 5 min.
3. Transfer dissociated cells from the flask to a 15-mL tube and spin cells at low speed (1500 rpm) and resuspend pellet in DMEM cell culture medium. Place a glass coverslip into each 35-mm plastic cell culture dish (Nunc), seed BHK cells, and incubate cells overnight.

3.2 Transfection

1. When BHK cells have reached 50–70% confluency, transiently transfect with Lipofectamine 3000 as per the manufacturer's instructions. Briefly, make two reaction mixtures. Reaction mixture A; 1.5 μg of total plasmid DNA (1 μg of EGFR-GFP and 0.5 μg of mKate-P2a-APEX-GBP) is added to 125 μL of Optimem (Gibco) with 3 μL of P3000 reagent. Vortex Reaction mixture A and incubate for 5 min. Reaction mixture B; 1.5 μL of Lipofectamine 3000 is added to 125 μL of Opti-mem and vortexed. The contents of Reaction mixture B are added to Reaction mixture A, vortexed, and the final solution is incubated 5 min after which the final solution is added to the cell media.
2. 5 h later, aspirate the DMEM cell culture medium/Lipofectamine 3000 mix and replace with fresh DMEM cell culture medium and leave the cells overnight at 37 °C (5% CO_2).

3.3 Expression Validation

1. Check transfection efficiency in a widefield fluorescence microscope. GFP fluorescence intensity indicates the expression of the EGFR-GFP construct of interest, and mKate fluorescence intensity indicates co-transfection of the APEX-GBP construct.
2. A confluency of 70–90% is ideal for EM visualization (Fig. 2a–c).

3.4 Ligand Stimulation

1. Prior to commencing ligand stimulation prepare fresh: 2.5% Glutaraldehyde in phosphate-buffered saline; 4% paraformaldehyde in phosphate-buffered saline.
2. Warm serum-free DMEM to 37 °C prior to serum starvation. Add serum-free medium to cells and incubate for 4 h prior to ligand stimulation.
3. Make two solutions for ligand stimulation pre-warmed to 37 °C. Solution 1: serum-free control solution containing no

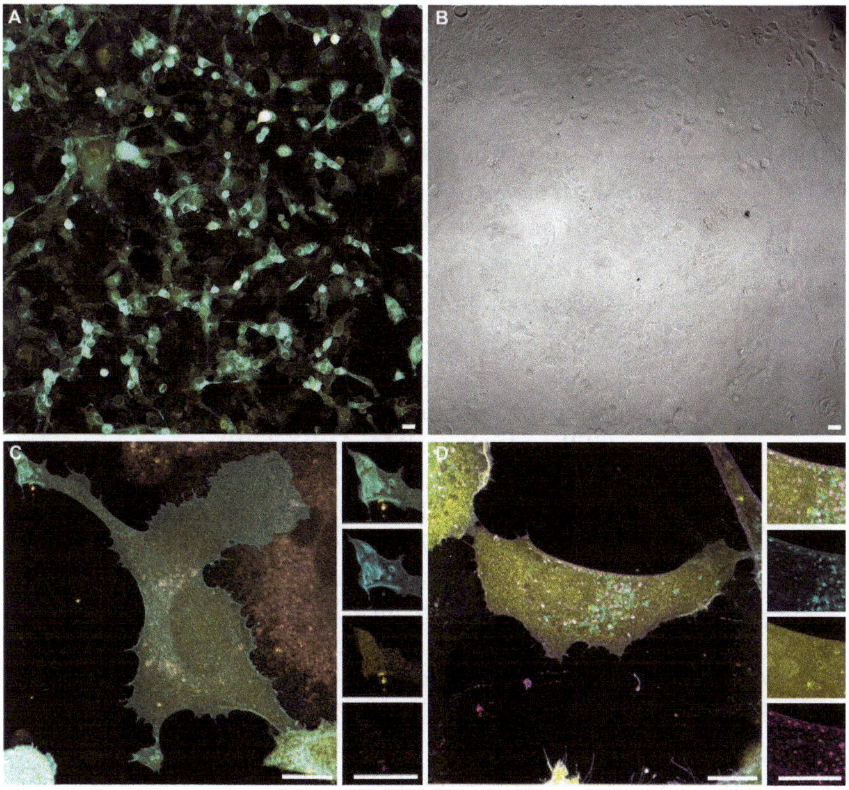

Fig. 2 Fluorescence validation of plasmid expression and EGF activation. (**a**) Low magnification widefield fluorescence image of BHK cells transfected with EGFR-GFP (green) and mKate-P2A-APEX-GBP (yellow). (**b**) Brightfield image demonstrated a confluency of 70–90%, which is ideal for EM processing. (**c**) Maximum intensity z-projection from a confocal z-stack of an unstimulated control BHK cell transfected with EGFR-GFP (cyan) and APEX-GFP (by mKate fluorescence; yellow) showing few internal puncta of EGFR-GFP in serum starved cells and an enrichment in plasma membrane ruffles (inset). (**d**) Maximum intensity z-projection from a confocal z-stack of an AF647-EGF (magenta) stimulated (15 min) BHK cell transfected with EGFR-GFP showing a reduction in plasma membrane EGFR with a commensurate increase in endosomal EGFR signal. Scale bars = 10 μm

ligand. Solution 2: serum-free DMEM solution containing AlexaFluor647-EGF (AF647-EGF) diluted to a final concentration of 10 ng/mL (*see* **Note 1**).

4. Treat dishes with control Solution 1 or EGF containing Solution 2 for 15 min at 37 °C in an incubator.

3.5 Fixation of Cells for Fluorescence Microscopy and Electron Microscopy

1. Wash each dish with pre-warmed (37 °C) phosphate-buffered saline.
2. Remove coverslips from each dish and transfer into 4% PFA in PBS for FlM validation of receptor internalization by EGF stimulation (*see* Subheading 3.6). Incubate for 30 min.

3. Add 2.5% glutaraldehyde in phosphate-buffered saline to the dishes in the fumehood for 1 h at room temperature to fix cells for EM (*see* Subheading 3.7).

3.6 Validation of EGF Stimulation and Receptor Internalization by Fluorescence Microscopy

1. Wash coverslips five times in PBS for 1 min.
2. Wash coverslips five times rapidly with water and mount with ProLong Glass Antifade on glass slides.
3. Leave slides for 24 h at room temperature to cure. Protect the slides from light.
4. Acquire whole cell z-stacks of the coverslips on a Zeiss LSM 900 with Airyscan 2 under the control of Zen Blue v3.6. Control non-stimulated BHK cells transfected with EGFR-GFP, and mKate-P2A-APEX-GBP demonstrate a predominant plasma membrane distribution enriched within plasma membrane ruffles EGFR-GFP with limited small intracellular puncta (Fig. 2a). AF647-EGF stimulated BHK cells transfected with EGFR-GFP, and mKate-P2A-APEX-GBP demonstrate an increase in intracellular EGFR-GFP signal with commensurate reduction in plasma membrane signal and abundant AF647-EGF within endosomal compartments (Fig. 2b) (*see* **Note 2**).

3.7 Electron Microscopy Processing for APEX Visualization

1. Prior to commencing EM processing prepare the following solutions fresh: 0.1-M sodium cacodylate buffer, 2.5% Glutaraldehyde in sodium cacodylate buffer; DAB in 0.1-M sodium cacodylate buffer; LX112 resin, 1% OsO_4 in 0.1 M sodium cacodylate buffer.
2. Wash cells twice in PBS for 1 min.
3. Wash dishes with 0.1-M sodium cacodylate buffer three times to ensure the removal of all glutaraldehyde from the dishes prior to DAB reaction.
4. Add 1 mg/mL DAB solution in 0.1-M sodium cacodylate buffer to the cells for 2 min.
5. Remove DAB wash solution and add fresh 1 mg/mL DAB solution in 0.1-M sodium cacodylate buffer supplemented with 10 mM H_2O_2 for 5–30 min. A visible brownish reaction product will develop over time (*see* **Note 3**).
6. Wash dishes five times in 0.1-M sodium cacodylate buffer to remove any free DAB.
7. Add 1% OsO_4 in 0.1 M sodium cacodylate buffer to the cells for 2 min. This short incubation is sufficient for preferential deposition of osmium atoms at the site of the DAB reaction product.
8. Wash cells in 0.1-M sodium cacodylate buffer three times for 30 s to remove remaining OsO_4 from the dish.

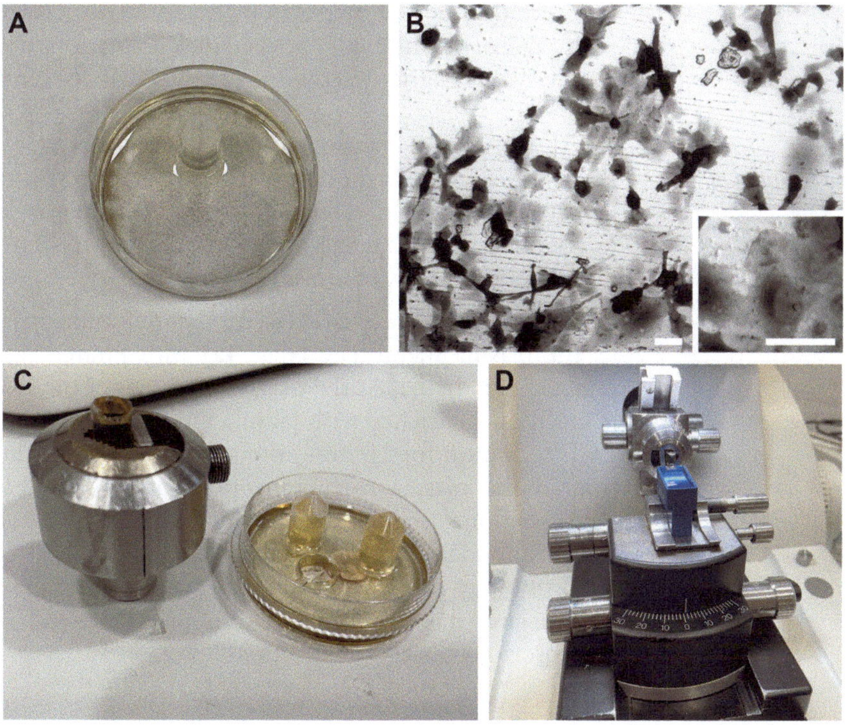

Fig. 3 APEX-GBP processing of EGFR-GFP expressing cells. (**a**) Flat embedding of EGFR-GFP transfected cells in LX112 resin with a BEEM capsule. (**b**) Brightfield imaging of an LX112 resin block post-polymerization. BHK cells have been transfected with EGFR-GFP and mKate-P2A-APEX-GBP and stimulated with AF647-EGF for 15 min. The cells demonstrate an accumulation of intracellular DAB reaction product. Scale = 40 μm. (**c**) Mounting BEEM capsules into the microtome chuck. (**d**) Diamond knife with water mounted on an ultramicrotome for ultrathin sectioning

9. Wash cells three times in H_2O to remove the remaining sodium cacodylate buffer.
10. All subsequent steps in the EM sample preparation are performed in a PELCO Biowave microwave processor (*see* **Note 4**).
11. Serial dehydration is performed with increasing percentages of ethanol in water in the Biowave for 40 s at 250 W per step. Steps are as follows; 30%, 50%, 70%, 90%, 100%, and 100% ethanol (*see* **Note 5**).
12. LX112 resin is serially infiltrated into cells with increasing percentages of LX112 resin in ethanol in the Biowave for 3 min at 250 W under vacuum. Steps are as follows: 33%, 66%, 100%, 100%, and 100% (*see* **Note 6**).
13. Polymerize dishes for 24 h at 60 °C in an oven. Allow the dishes to cool to room temperature, use pliers to separate the BEEM capsule from the bulk resin and separate the embedded cells from the base of the dish (Fig. 3b).

14. Cut ultrathin 60–70 nm sections on an ultramicrotome fitted with a diamond knife (Fig. 3c) and pick up sections with carbon coated formvar grids. Allow the grids to dry (*see* **Note 7**).

3.8 Electron Microscopy Imaging of EGFR-GFP with Modular APEX-GBP

1. Insert grids into a transmission electron microscope at 80 kV. Find the focus of the sections and at low magnification (>1000× magnification) search for cells with slight differences in contrast. Insert a small objective aperture to better resolve differences in APEX contrast (*see* **Note 8**).
2. In serum starved cells EGFR-GFP is predominantly localized to the plasma membrane (Fig. 4a) with limited intracellular puncta (Fig. 4a, b). At the plasma membrane, the EGFR-GFP is specifically enriched with plasma membrane ruffles (Fig. 4c). EGF activation for 15 min prior to fixation results in a reduction in the plasma membrane pool of EGFR-GFP and an increase in intracellularly localized receptor (Fig. 4a). EGFR-GFP is abundant in endosomal membrane compartments with a morphology consistent with early-stage endosomes (Fig. 4b) and in later-stage multivesicular bodies (Fig. 4c) consistent with previous observations.

4 Notes

1. EGF concentrations vary between tissues, and the local concentration of EGF affects the endocytosis of the receptor [20]. High concentrations of EGF (above 10 ng/mL) are found in fluids such as milk and saliva [21, 22] and may be generated at local sites during wound healing [23]. EGFR stimulation at these kinds of concentrations can result in both clathrin-mediated and clathrin-independent endocytosis of the receptor [16]. Lower concentrations of EGF approaching 1–2 ng/mL are present in mammalian tissues and result in entry of the receptor predominantly through clathrin-mediated endocytosis [20].
2. Low level of expression for both EGFR-GFP and mKate-P2A-APEX-GBP is preferred. Low EGFR-GFP expression is required to minimize overexpression artefacts. To avoid these artefacts, we have previously utilized genome editing to detect proteins at endogenous levels using a particulate APEX reaction product [12]. Other groups have demonstrated the utility of CRISPR/CAS9 genome editing to study the trafficking of the EGFR [24]. The P2A cleavage site allows for a direct inference of APEX-GBP expression in each cell through the fluorescence intensity of the mKate reporter. In high expressing cells, there can be a saturation of available GFP epitopes by overexpression of the mKate-P2A-APEX-GBP, which can in

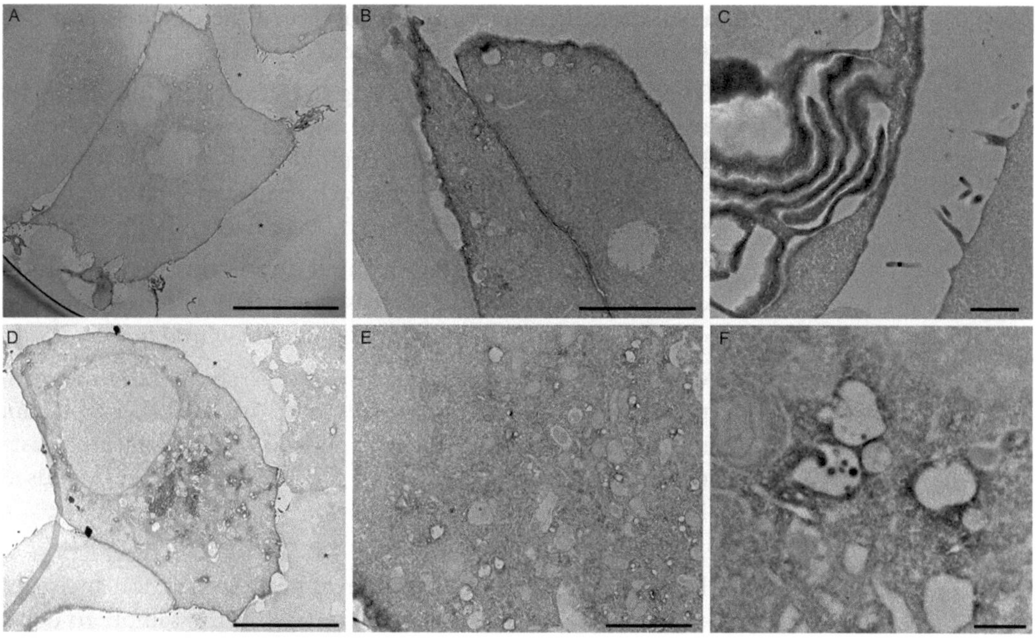

Fig. 4 APEX-GBP detection of EGFR-GFP distribution in a transmission electron microscope. (**a–c**) Control serum starved cells. (**a**) Low magnification electron micrograph of BHK cells expressing EGFR-GFP and mKate-P2A-APEX-GBP after the DAB reaction and contrasting with OsO_4. EGFR-GFP is high enriched at the plasma membrane and within plasma membrane ruffles. Untransfected internal control cells highlighted by *. Scale = 20 μm. Higher-magnification electron micrographs showing that EGFR-GFP is tightly restricted to the plasma membrane in serum starved cells. Scale = 2 μm. (**c**) EGFR-GFP is highly enriched within plasma membrane ruffles and filopodia. Scale = 1 μm. (**d–f**) AF647-EGF stimulated cells (15 min). (**d**) Ligand stimulation results in redistribution of EGFR from the plasma membrane to intracellular compartments. Untransfected internal control cells highlighted by *. Scale = 10 μm. (**e**) Intracellular compartments decorated with EGFR-GFP and APEX-GBP resemble an endosomal distribution. Scale = 1 μm. (**f**) High magnification electron micrograph showing EGFR-GFP and APEX-GFP on the cytoplasmic face of endosomes and accumulating within intralumenal vesicles. Scale = 500 nm

turn increase the free APEX-GBP cytoplasmic background. This problem can be overcome by two approaches: (i) strict limitation of the expression level of the mKate-P2A-APEX-GBP construct (as used here) or (ii) the use of the conditionally stabilized mKate-P2A-APEX-csGBP construct (Addgene #108875). The conditional stabilization of the APEX-GBP is achieved through additional point mutations within the nanobody domain such that, if unbound, the nanobody domain of the APEX-GBP is targeted for proteasomal degradation [10]. This construct, therefore, allows for sensitive detection of a GFP-tagged protein of interest even when APEX-GBP expression is in excess.

3. DAB is a diffusing reaction product. Longer development times will increase the signal, but this will limit the attainable

resolution. It is important to optimize the development time for the expression of your protein of interest.

4. The following sample processing steps can be performed on the bench and without the use of the microwave processor, but longer incubation times are required. Each ethanol step should be 10 min as a minimum, and each resin step should be 60 min with the final infiltration step to be an overnight infiltration.
5. Solution transfers need to be rapid during this step as cells can dry out quickly with the increased volatility of ethanol. Drying of cells will impact the quality of ultrastructural preservation.
6. Cells are flat embedded. For ease of separation of the cells from the dish, place a BEEM capsule filled with 100% LX112 and invert onto the dish (Fig. 3a). LX112 is required if using plastic tissue culture dishes; however, if a glass bottom dish is used, Epon resin is a suitable alternative.
7. Post-staining is not required for APEX-GBP detection of EGFR-GFP protein distribution. The DAB reaction and osmication steps provide sufficient contrast in the sample to determine distribution without the addition of uranyl acetate or Reynold's lead citrate.
8. Interaction between heavy atoms (e.g., the osmium at the site of the APEX-tag) and the electron beam at lower accelerating voltage results in greater scattering of electrons and increased amplitude contrast. By combining lower kV imaging with a smaller objective aperture there will be greater contrast to the sample. This increase in contrast will simplify the determination of APEX reaction product in transfected cells from background.

References

1. Ariotti N, Wu Y, Okano S et al (2021) An inverted CAV1 (caveolin 1) topology defines novel autophagy-dependent exosome secretion from prostate cancer cells. Autophagy 17(9): 2200–2216. https://doi.org/10.1080/15548627.2020.1820787
2. Follett J, Bugarcic A, Yang Z et al (2016) Parkinson disease-linked Vps35 R524W mutation impairs the endosomal association of retromer and induces alpha-synuclein aggregation. J Biol Chem 291(35):18283–18298. https://doi.org/10.1074/jbc.M115.703157
3. Dunn KW, Kamocka MM, McDonald JH (2011) A practical guide to evaluating colocalization in biological microscopy. Am J Physiol-Cell Ph 300(4):C723–C742. https://doi.org/10.1152/ajpcell.00462.2010
4. Slot JW, Geuze HJ (2007) Cryosectioning and immunolabeling. Nat Protoc 2(10): 2480–2491. https://doi.org/10.1038/nprot.2007.365
5. Martell JD, Deerinck TJ, Sancak Y et al (2012) Engineered ascorbate peroxidase as a genetically encoded reporter for electron microscopy. Nat Biotechnol 30(11):1143. https://doi.org/10.1038/nbt.2375
6. Lam SS, Martell JD, Kamer KJ et al (2015) Directed evolution of APEX2 for electron microscopy and proximity labeling. Nat Methods 12(1):51–54. https://doi.org/10.1038/nmeth.3179
7. Pouchelon G, Vergara J, McMahon J et al (2022) A versatile viral toolkit for functional discovery in the nervous system. Cell Rep

Methods 2(6). https://doi.org/10.1016/j.crmeth.2022.100225
8. Li H, He CW, Zhu J et al (2022) A validated set of ascorbate peroxidase-based organelle markers for electron microscopy of Saccharomyces cerevisiae. mSphere 7(4):e0010722. https://doi.org/10.1128/msphere.00107-22
9. Ariotti N, Hall TE, Rae J et al (2015) Modular detection of GFP-labeled proteins for rapid screening by electron microscopy in cells and organisms. Dev Cell 35(4):513–525. https://doi.org/10.1016/j.devcel.2015.10.016
10. Ariotti N, Rae J, Giles N et al (2018) Ultrastructural localisation of protein interactions using conditionally stable nanobodies. PLoS Biol 16(4):e2005473. https://doi.org/10.1371/journal.pbio.2005473
11. Ariotti N, Hall TE, Parton RG (2017) Correlative light and electron microscopic detection of GFP-labeled proteins using modular APEX. Methods Cell Biol 140:105–121. https://doi.org/10.1016/bs.mcb.2017.03.002
12. Rae J, Ferguson C, Ariotti N et al (2021) A robust method for particulate detection of a genetic tag for 3D electron microscopy. elife 10. https://doi.org/10.7554/eLife.64630
13. Caldieri G, Malabarba MG, Di Fiore PP et al (2018) EGFR trafficking in physiology and cancer. Prog Mol Subcell Biol 57:235–272. https://doi.org/10.1007/978-3-319-96704-2_9
14. Sibilia M, Wagner EF (1995) Strain-dependent epithelial defects in mice lacking the EGF receptor. Science 269(5221):234–238. https://doi.org/10.1126/science.7618085
15. Lemmon MA, Schlessinger J (2010) Cell signaling by receptor tyrosine kinases. Cell 141(7):1117–1134. https://doi.org/10.1016/j.cell.2010.06.011
16. Goh LK, Huang F, Kim W et al (2010) Multiple mechanisms collectively regulate clathrin-mediated endocytosis of the epidermal growth factor receptor. J Cell Biol 189(5):871–883. https://doi.org/10.1083/jcb.201001008
17. Henriksen L, Grandal MV, Knudsen SL et al (2013) Internalization mechanisms of the epidermal growth factor receptor after activation with different ligands. PLoS One 8(3):e58148. https://doi.org/10.1371/journal.pone.0058148
18. Roepstorff K, Grandal MV, Henriksen L et al (2009) Differential effects of EGFR ligands on endocytic sorting of the receptor. Traffic 10(8):1115–1127. https://doi.org/10.1111/j.1600-0854.2009.00943.x
19. Carter RE, Sorkin A (1998) Endocytosis of functional epidermal growth factor receptor-green fluorescent protein chimera. J Biol Chem 273(52):35000–35007. https://doi.org/10.1074/jbc.273.52.35000
20. Sorkin A, Goh LK (2009) Endocytosis and intracellular trafficking of ErbBs. Exp Cell Res 315(4):683–696. https://doi.org/10.1016/j.yexcr.2008.07.029
21. Connolly JM, Rose DP (1988) Epidermal growth factor-like proteins in breast fluid and human-milk. Life Sci 42(18):1751–1756. https://doi.org/10.1016/0024-3205(88)90041-0
22. Murdoch-Kinch CA, Russo N, Griffith S et al (2011) Recovery of salivary epidermal growth factor in parotid saliva following parotid sparing radiation therapy: a proof-of-principle study. Oral Surg Oral Med Oral Pathol Oral Radiol Endod 111(1):64–70. https://doi.org/10.1016/j.tripleo.2010.09.005
23. Schultz G, Rotatori DS, Clark W (1991) Egf and Tgf-alpha in wound-healing and repair. J Cell Biochem 45(4):346–352. https://doi.org/10.1002/jcb.240450407
24. Larsen MB, Verdaguer MP, Schmidt BF et al (2019) Generation of endogenous pH-sensitive EGF receptor and its application in high-throughput screening for proteins involved in clathrin-mediated endocytosis. elife 8:e46135. https://doi.org/10.7554/eLife.46135

Chapter 8

Some Guiding Principles for a "Simple" Correlative Light Electron Microscopy Experiment

Elina Mäntylä and Paul Verkade

Abstract

In recent years, Correlative Multimodal Imaging (CMI) has become an "en vogue" technique and a bit of a buzzword. It entails combining information from different imaging modalities to extract more information from a sample that would otherwise not be possible from each individual technique. The best established CMI technology is correlative light and electron microscopy (CLEM), which applies light and electron microscopy on the exact same sample/structure. In general, it entails the detection of fluorescently tagged proteins or structures by light microscopy and subsequently their relative intracellular localization is determined with nanometer resolution using transmission electron microscopy (TEM). Here, we describe the different steps involved in a "simple" CLEM approach. We describe the overall workflow, instrumentation, and basic principles of sample preparation for a CLEM experiment exploiting stable expression of fluorescent proteins.

Key words Multimodal imaging, CLEM, Sample preparation, Immunogold labelling, Fluorescence microscopy, EM correlation

1 Introduction

It has been well documented that Correlative and Multimodal Imaging (CMI) approaches are a very powerful tool in the pallet of research techniques available to researchers to address a wide range of biological questions at a technically very challenging level [e.g., 1–3]. In these examples, the biological question required the researchers to apply specific technologies to answer their question. Any CMI experiment will also always have to make a compromise between parameters such as temporal and spatial resolution, volume, throughput, and the native state of the sample [4]. One of the leading principles should always be to keep the experimental approach as simple as possible.

One of the simpler CMI technologies is correlative light and electron microscopy (CLEM) of resin-embedded samples. CLEM empowers researchers to study the intracellular structures, protein

Christoph Wülfing and Robert F. Murphy (eds.), *Imaging Cell Signaling*, Methods in Molecular Biology, vol. 2800, https://doi.org/10.1007/978-1-0716-3834-7_8,

interactions, and relative locations of fluorescently tagged proteins in a nanoscopic resolution by correlating information acquired with fluorescence microscopy (FM) and electron microscopy (EM) [5]. In this method, the FM indicates the localization of one or more fluorescent molecules of interest, while the EM images provide the corresponding ultrastructural context. For this, cells can be grown on micropatterns and fixed for visualization. After fixation, endogenously expressed proteins must be immunolabelled before imaging, but stable fluorescent protein-expressing cells can be directly visualized with a laser scanning confocal microscope (LSCM) and registered for further survey by using the grid patterns. First studies exploiting stable expression of fluorescent proteins in CLEM were performed in 2000, where membrane trafficking of GFP-tagged vesicles was visualized [6–8]. For CLEM, the use of fluorescent protein tags shortened sample treatment times and aided in avoiding antibody background-related artifacts, which generally lead to hampered signal-to-background ratios. However, the loss of fluorescence in CLEM following the fixations and embedding poses difficulties in imaging. To overcome this problem and to identify the targets of interest, nanogold labelling is still often used after the LSCM imaging [e.g., 9].

One of the breakthrough innovations applied in the Polishchuk and Mironov papers was the use of glass-bottom imaging dishes where the glass-bottom coverslip had a finder pattern embossed on it.

In CLEM workflows using such finder dishes, GFP-expressing cells are generally imaged under the light microscope before fixation, whereas endogenous proteins need to be detected by antibodies and are first fixed, immunolabelled, and then analyzed under the light microscope. Samples are then processed for electron microscopy. The embedded samples are then trimmed to an area that contains the region/cell(s) of interest, and ultrathin sections are cut. Following locating the same cells with the help of the grid pattern, the samples are imaged at high resolution with electron microscopy (EM).

Virtually, each CLEM experiment will follow this same set of steps, starting with the biological sample, followed by light microscopy, processing, electron microscopy, and ending with the analysis (Fig. 1).

Here, we describe a general CLEM workflow using HEK293 cells stably expressing Galectin8-GFP as an example. We explain how to use the finder dishes and highlight some of the technical aspects of the process.

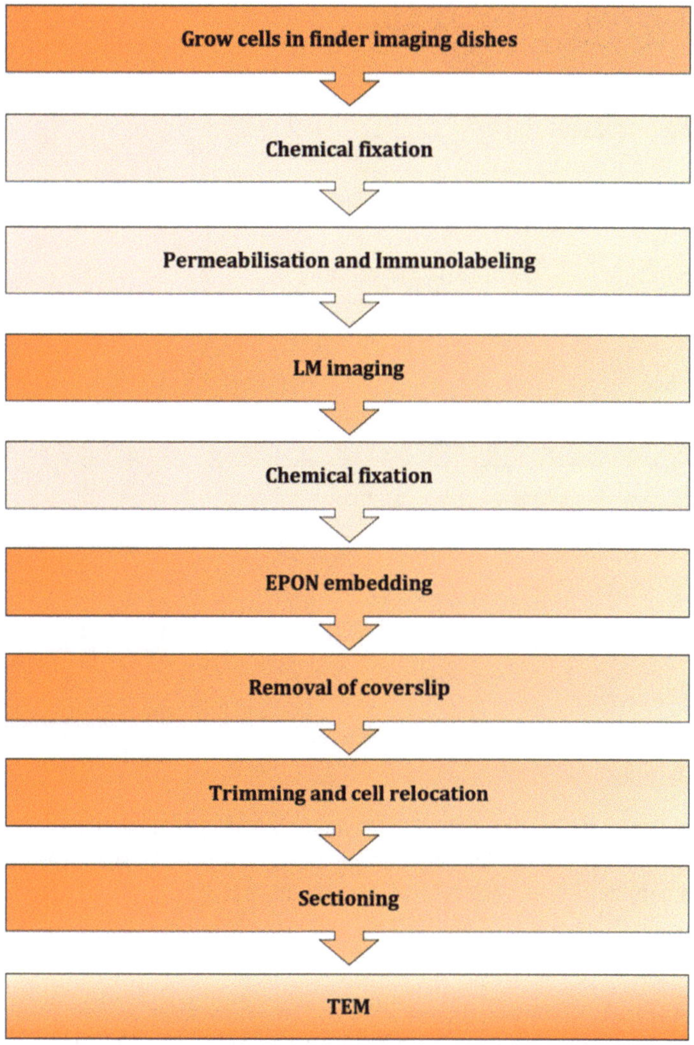

Fig. 1 Schematic overview of a CLEM workflow

2 Materials

2.1 Cell Culture and Fixation

1. Glass-bottom cell imaging dishes with gridded patterns (Mat-Tek, P35G-1.5-14-CGRD) (Fig. 2).
2. Fixative: 4% Paraformaldehyde in PBS.

2.2 Confocal Laser Scanning Microscopy

1. SP8 confocal laser scanning microscope, Leica Microsystems.

2.3 TEM Sample Processing

1. Ultra microtome UC7, Leica Microsystems.
2. Diamond knife, 45°, Diatome.
3. Transmission electron microscope, Tecnai12 120 kV BioTwin Spirit TEM with a FEI CETA 16 M CMOS camera.

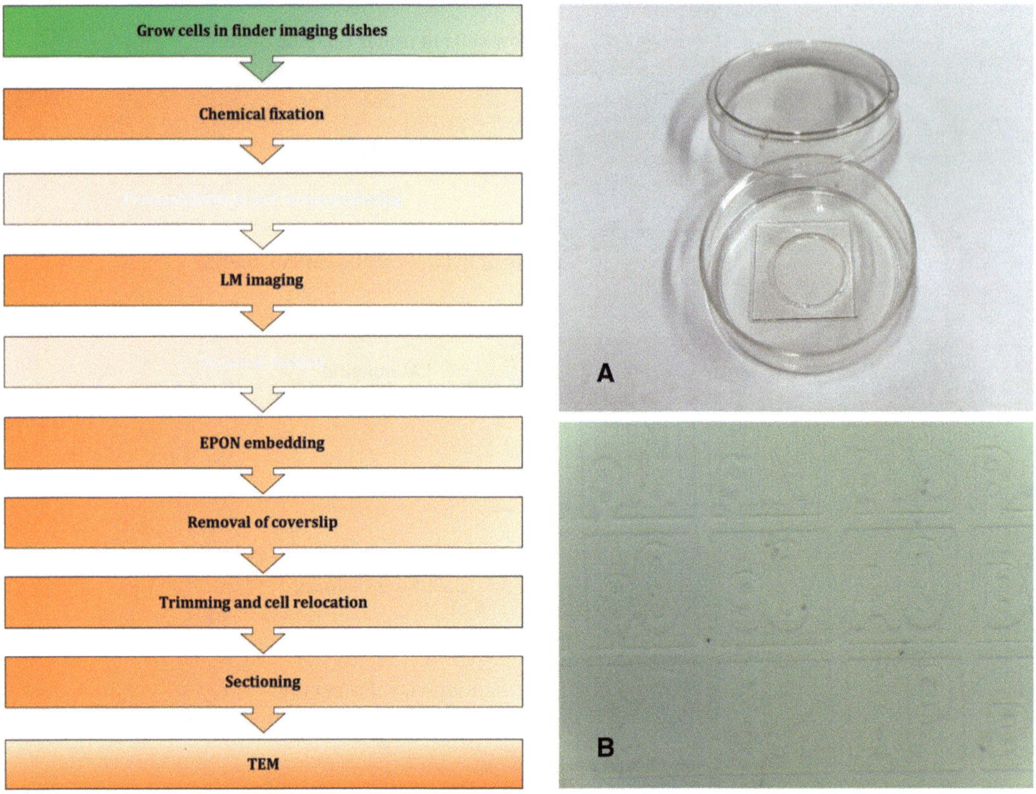

Fig. 2 Schematic overview of a CLEM workflow highlighting the finder imaging dishes. The finder dish looks like an ordinary live cell imaging dish (**a**) and can be used as such. The coverslip is embossed with a finder pattern (**b**) that allows relocation of areas of interest

4. Glutaraldehyde (Polysciences, 111-30-8).
5. Silver enhancement kit (SE-EM) (Aurion, 500.033).
6. Osmium tetroxide (OsO4, Merck, O5500).
7. Uranyl acetate (UA, old source from Sigma Aldrich).
8. Epon 812 mix: 96 g epon 812 resin, 64 g DDSA (dodecenyl-succinic anhydride), 40 g MNA (mono nadic anhydride), 6 g BDMA (benzyldimethylamine).

3 Methods

In our example, we will be seeding HEK293 cells stably expressing Galectin8-GFP in the finder imaging dishes, working our way through the workflow to end up with our sample as thin resin-embedded sections in a transmission EM. For cells that will be transiently transfected with a GFP construct or where an endogenous protein needs to be detected, procedures may be slightly different (see e.g [9]).

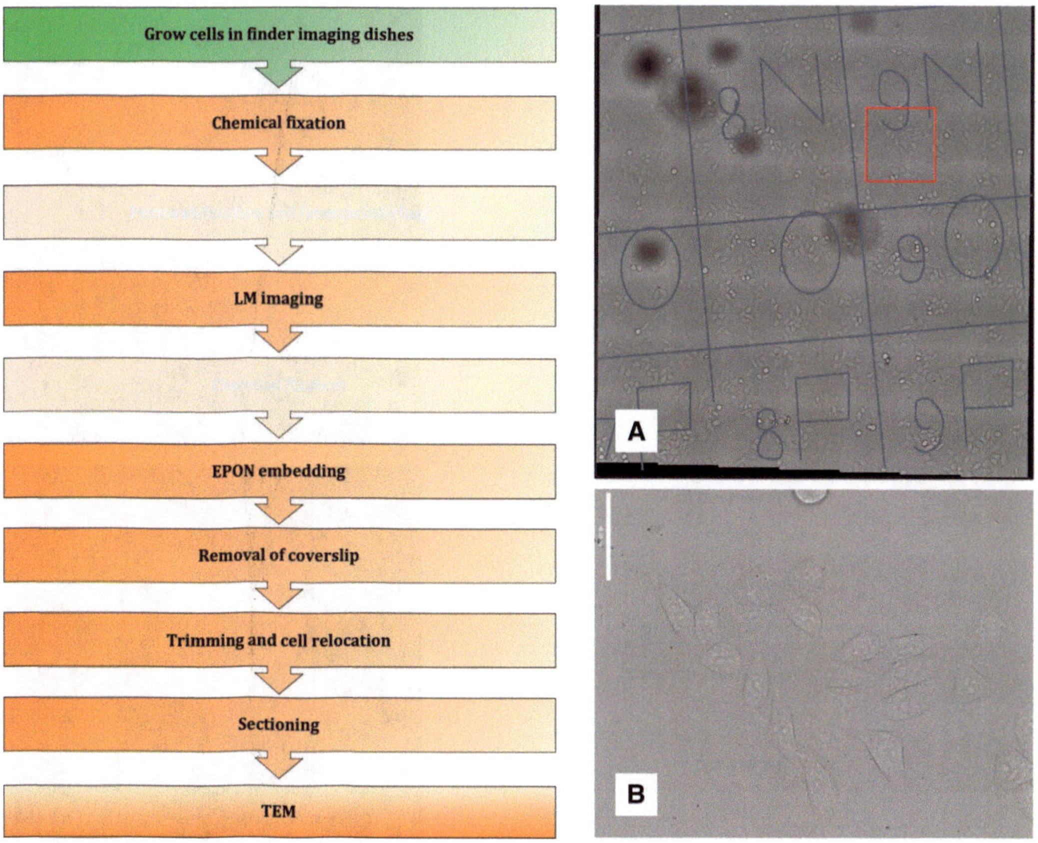

Fig. 3 Schematic overview of a CLEM workflow highlighting the seeding density of cells in the imaging dish. Cells should be used while they are still present as single cells or as small groups of cells (**a**). The brightfield image allows the visualization of the finder pattern (**a**) as well as detailed orientation of individual cells in a group (**b**)

3.1 Cell Culture and Fixation

1. Seed cells into the imaging dishes and leave overnight (*see* **Note 1**) (Fig. 3).
2. Fix the cells with paraformaldehyde in PBS for 15 min (*see* **Note 2**).

3.2 Cell Culture, Light Microscopical Imaging and Fixation

1. Image the cells in PBS with an inverted confocal laser scanning microscope.
2. Identify the cells/region of interest (ROI) with the aid of the grid.
3. Image ROI first by taking an overview image. This will aid in pattern recognition later and retracing the cell and organelle of interest (Fig. 4a).
4. Acquire higher magnification images (Fig. 4b). This can be done by digital Zoom or switching to a higher magnification objective (*see* **Note 3**).

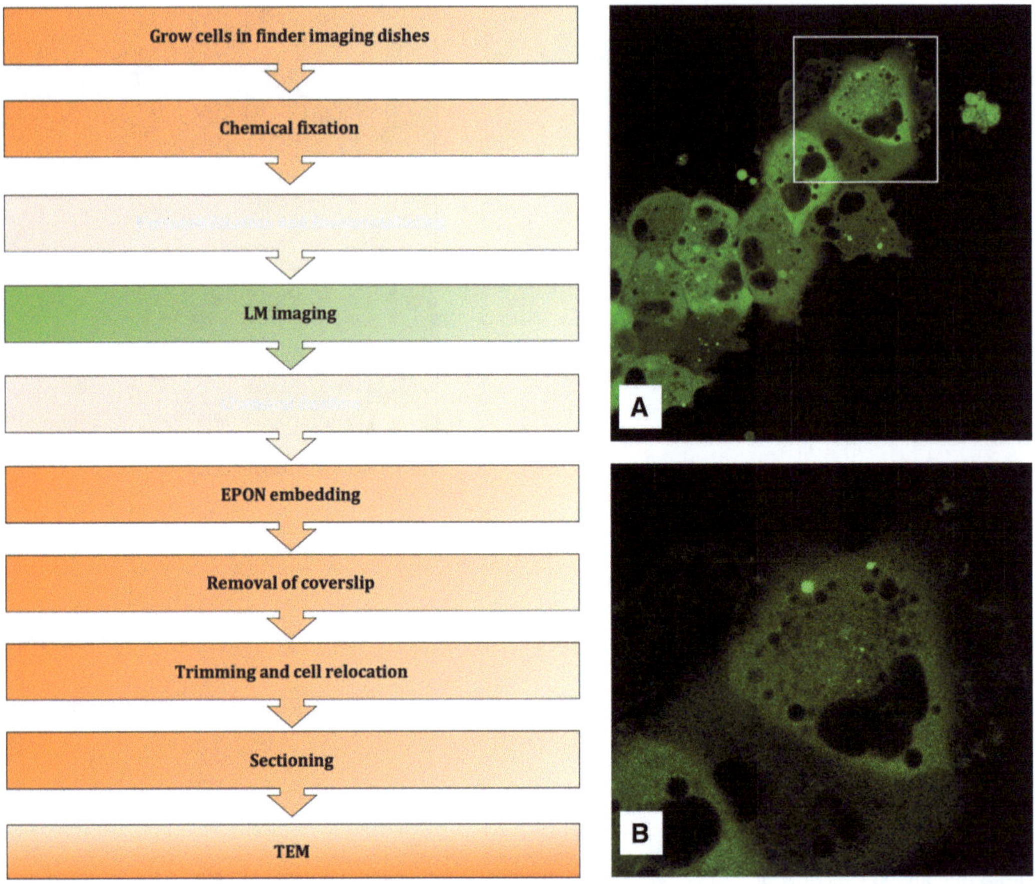

Fig. 4 Schematic overview of a CLEM workflow highlighting the light microscopy imaging step. Following acquisition of a bright field overview image (Fig. 3) fluorescence imaging captures the overall distribution of GFP-galectin-8 in the cells (**a**) and inside the cytoplasm (**b**) of expressing cells and accumulated in fluorescent spots. Boxed area in (**a**) is enlarged in (**b**)

3.3 Electron Microscopy Processing

1. Following light microscopy imaging, the samples are processed for electron microscopy. A secondary (stronger) fixation may help to preserve the ultrastructure better. Processing for electron microscopy has been described in detail in other papers [e.g., 6, 9, 10] but generally follows the following steps. We refer to the literature for detailed protocols but would like to note that there are many variations on the protocol and each will be slightly different (it is sometimes said that if there are 100 electron microscopists, there are 101 different protocols).
2. Post-fixation.
3. Staining.
4. Osmication.
5. Uranyl staining.
6. Dehydration.

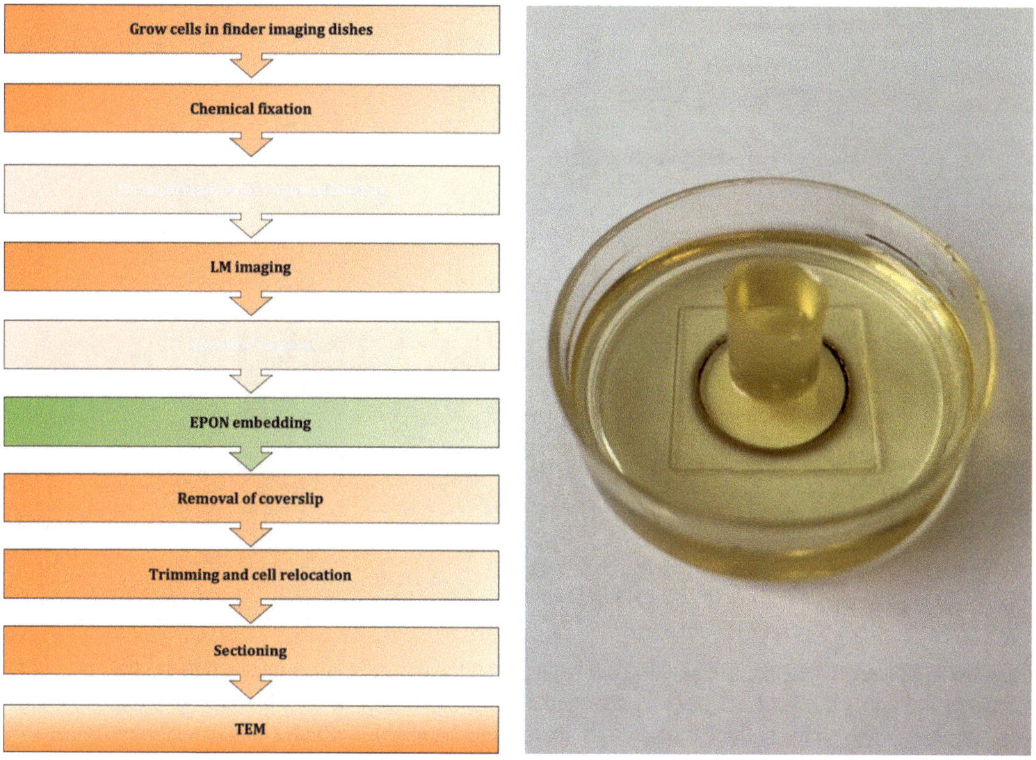

Fig. 5 Schematic overview of a CLEM workflow highlighting the end product of the EM sample processing and resin infiltration. The imaging dish is filled with a thin layer of resin that has been hardened and a stub is added to the resin in the middle

7. infiltration with resin.
8. Curation of the resin.
9. Attach a premade polymerized Epon stab to the top of the Epon-embedded sample block with a drop of fresh Epon and cure in the oven at 60 °C overnight (previously described in [11]). At the end of the curing, one should end up with a dish with a "handle" (Fig. 5) (*see* **Note 4**).

3.4 Trimming

1. As the surface area of the coverslip, now covered with polymerized resin, is way too big to be sectioned, the block face should be trimmed down as small as possible (*see* **Note 5**).
2. Remove the glass coverslip and plastic dish [11].
3. Fix the Epon block in a metal specimen holder designed for ultramicrotomes (Fig. 6a and b).
4. Inspect the Epon block surface under a stereo microscope for relocation of the cells and use a razor blade to gently mark the ROI (Fig. 6c) (*see* **Note 6**).

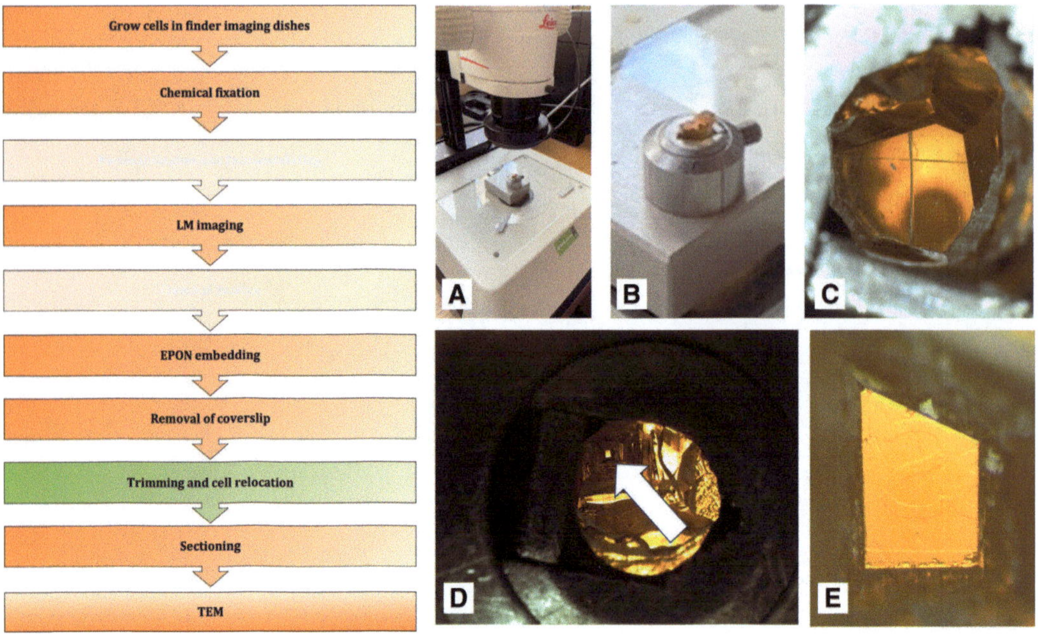

Fig. 6 Schematic overview of a CLEM workflow highlighting the different steps involved in the trimming of the sample to the right size. First the block is clamped in a special holder (**a** with **b** showing a zoomed image). Marks are scratched on the surface the serve as approximate indicators, which areas can be removed (**c**). The sample is trimmed down to a very small size (**d**, arrow indicating the shiny surface of the surface). (**e**) shows the zoom in of (**d**). The letter and number; S and part of the 8 are still left for reference. Individual cells can be recognized as dark spots (arrow)

5. Under the stereo microscope, manually trim the sides of the Epon block with a razor blade (previously described in [11]). Gradually work the block to smaller areas (Fig. 6d) (*see* **Note 7**).
6. After trimming, the ROI should be in the middle of the block, ideally showing some of the finder pattern (Fig. 6e).

3.5 Sectioning

1. The block is sectioned, and sections are collected on slot grids (Fig. 7) according to standard procedures, which will not be described here (*see* **Notes 8** and **9**).

3.6 EM Analysis

1. Once the grids are ready, it is time to go to the electron microscope and analyze the data. We prefer to take a laptop with our light microscopy data along so we can correlate our data on the spot and zoom into the regions of interest. Now, it is important that we have chosen our area of interest in a place that is easy to recognize, as highlighted in Subheading 3.1:

   A single cell.

   A small group of cells (up to five cells).

   At the edge of a larger group of cells.

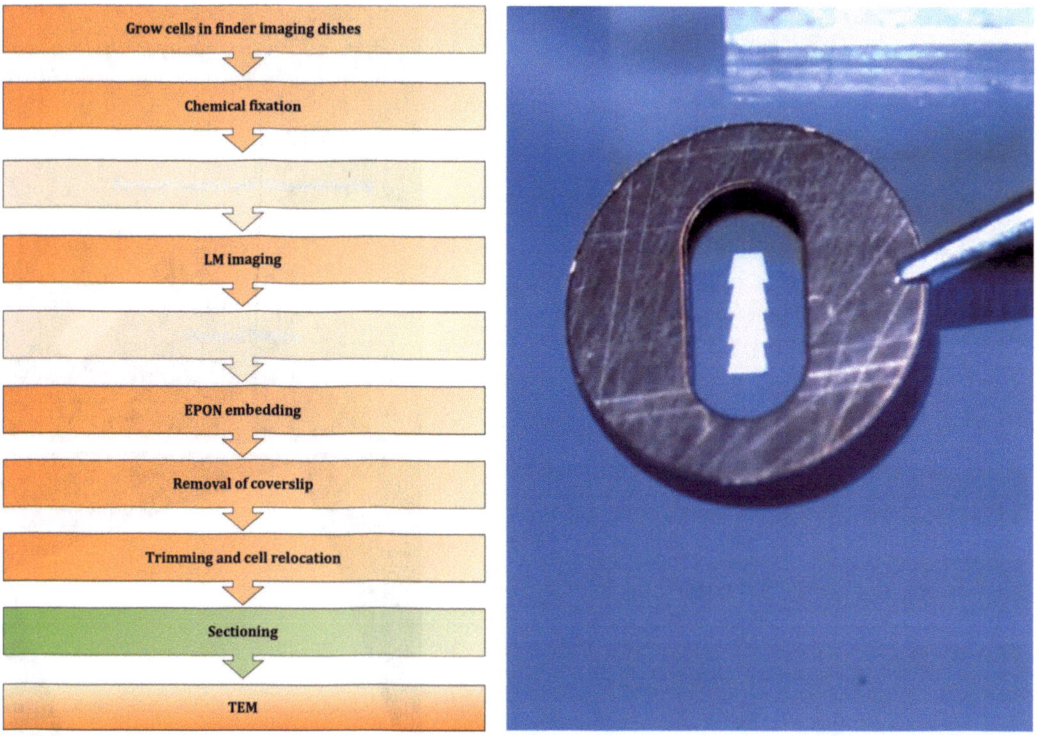

Fig. 7 Schematic overview of a CLEM workflow highlighting the sectioning step. Following alignment of the block, serial sections can be collected on a slot grid with a pioloform. This step shows the importance of using slot grids as bars of a Mesh grid would potentially cover the area of interest

2. Here, we have used a cell that is on the edge of a large cluster of cells (Fig. 8a).
3. Relocate the larger area of interest in the electron microscope using the overview images from the light microscope (*see* **Notes 10** and **11**).

3.7 Final Correlation

1. Once the right orientation is assured, one can begin the final correlation and overlay of the data.
2. One can use dedicated software to make those alignments. Software solutions that do this include ec-CLEM [12], CLEM-reg [13] or custom-made solutions [14], generally with a precision in the range of 50 nm. In our example, we have additional "high-resolution" landmarks in our sample that facilitate precise alignment of the two datasets, which allowed us to obtain a high-precision correlation without the need of additional software.
3. Cells contain many endocytic structures (endosomes, multivesicular bodies, and lysosomes) that exclude any cytosolic proteins. GFP expression of a cytosolic membrane-associated protein such as the galectin-8 in our example will therefore be

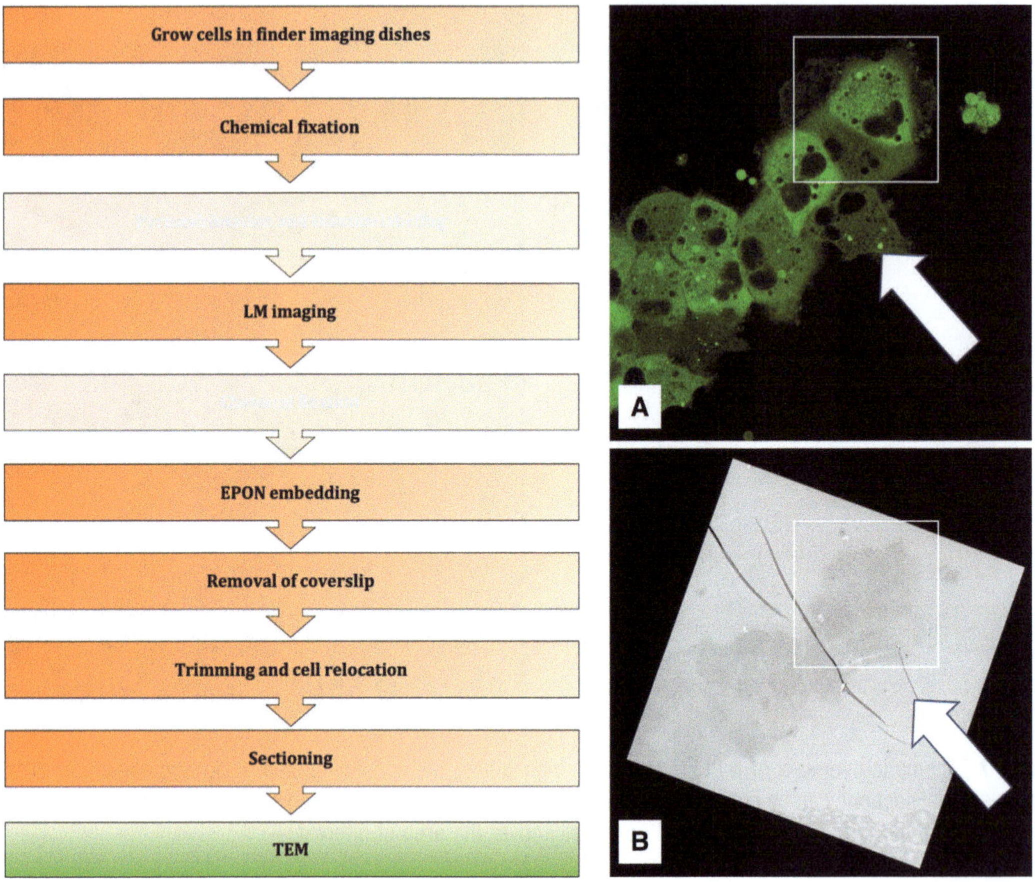

Fig. 8 Schematic overview of a CLEM workflow highlighting the initial correlation step at the electron microscope. Overview images of the fluorescence light microscopy can now be correlated to the cellular ultrastructure down the electron microscope

completely void of fluorescence in the endocytic structures, rendering them "black" inside the cells (see, e.g., Fig. 9b). Although not the case for all GFP-expressing constructs, our protein is also completely absent from the nucleus, leaving a big black hole in the cell (Fig. 9b).

4. With these landmarks in place, we can now zoom into our area of interest both in the light microscopy data as well as the electron microscopy data (Fig. 9). This allowed us to identify the bright fluorescent "blob" in the light microscopy as a tubular membrane organelle in the electron microscope. Here, we will not go into the biological question, which is always paramount when performing a (CLEM) experiment [5].

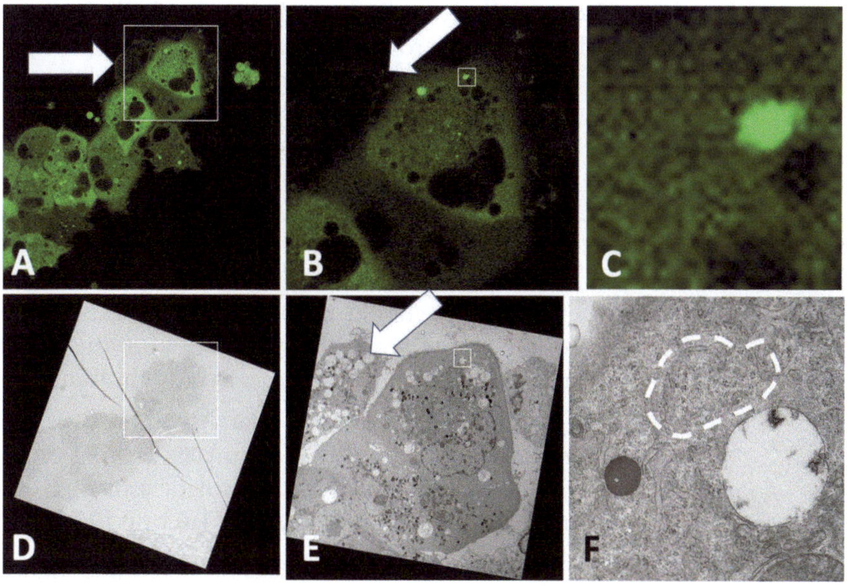

Fig. 9 Final outcome of the CLEM experiment shows that one can zoom in from a group of cells to the individual cell and finally to the structure of interest both in the light (**a**–**c**) and the electron (**d**–**f**) microscopy. In the final step, the fluorescent "blob" highlighted in the boxed area in (**b**) and in (**c**) can be identified as a tubulo-membranous structure (highlighted area in **f**). Note that very low expressing cells (arrow in **a** and **b**) are now also visualized in the electron microscope

3.8 Conclusion

1. We have described the main steps in a "simple" CLEM experimental workflow. It uses standard live cell imaging dishes with a slight modification; an embossed finder pattern. This workflow allows an easy introduction into the world of CLEM technologies and can generate very impactful results (e.g., [6, 9]). Scientists with a live cell imaging background will be very familiar with the concept of imaging dishes, while electron microscopists will know how to process the samples for EM from the fixation step. By working together and a pair of steady hands for the exact trimming steps, most scientists with access to both a light and electron microscopy facility will be able to use this CLEM workflow in their science. We have focused here on the exact trimming of the sample and highlighted some steps using photos from actual experiments.
2. Many of the other steps have been highlighted in other chapters of Methods books (e.g. [11, 15). Those have been part of a series of Methods in Cell Biology books on Correlative Light Electron Microscopy [16–19], each describing a wide range of correlative microscopy approaches.
3. The technology in the CLEM field is still developing further [20, 21].

4. Also, the incorporation of other technologies apart from light and electron microscopy techniques is on the rise in so-called Correlative Multimodal Imaging (CMI) workflows (see, e.g., [2–4]). Such CMI approaches will open up new opportunities for scientists to obtain a holistic view of their sample.

4 Notes

1. Seeding density should be such that at the time of the light microscopy, single cells or small groups are the prevalent conformation of the cells (Fig. 3, *see* also Subheading 3.4).
2. The galectin-8 structures are relatively stable over time, so live imaging is not required. For organelles moving on the second scale or faster, live imaging can provide a history of the event and may aid identifying a structure later on [22, 23].
3. By switching objectives, there is the danger of moving the sample and hence not imaging the same cell at higher magnification.
4. One can also add a stub at the end of the resin infiltration and harden the stub with the resin. This can be more stable but can create air bubbles.
5. A steady hand is required, and it may happen that occasionally the cell of interest is accidentally trimmed away.
6. The cells are now located at the bottom of the Epon-embedded sample, and the grid pattern will appear mirrored.
7. This step can also be performed directly on an ultramicrotome.
8. If part of the finder pattern is left, the first sections will contain this and may aid orientation of the sample.
9. One can use Mesh grids, but this has the danger of the cell of interest being mask by one of the bars.
10. The section in the microscope may not only be rotated in relation to the LM image but also inverted (flipped) [15].
11. Due to the processing and the way cells grow on a support, one may lose cells observed in the LM compared to the EM part (Fig. 8).

Acknowledgments

We would like to thank the Wolfson Bioimaging Facility at the University of Bristol for their expert technical support. This work was funded by the Academy of Finland under the award number 332615 to Elina Mäntylä.

References

1. Hoffman DP, Shtengel G, Xu CS et al (2020) Correlative three-dimensional super-resolution and block-face electron microscopy of whole vitreously frozen cells. Science 367:eaaz5357. https://doi.org/10.1126/science.aaz5357
2. Walter A, Paul-Gilloteaux P, Plochberger B et al (2020) Correlated multimodal imaging in life sciences: expanding the biomedical horizon. Front Phys 8:47. https://doi.org/10.3389/fphy.2020.00047
3. Johnson BE, Creason AL, Stommel JM et al (2022) An omic and multidimensional spatial atlas from serial biopsies of an evolving metastatic breast cancer. Cell Rep Med 3:100525. https://doi.org/10.1016/j.xcrm.2022.100525
4. Scher N, Avinoam O (2021) 50 shades of CLEM: how to choose the right approach for you. Methods Cell Biol 162:1–11. https://doi.org/10.1016/bs.mcb.2020.08.001
5. Verkade P (2021) Correlative light electron microscopy. In: Imaging modalities for biological and preclinical research: a compendium, Vol 2. Ch 16, III.1.a-1 to III.1.a-7. https://doi.org/10.1088/978-0-7503-3747-2ch16
6. Polishchuk RS, Polishchuk EV, Marra P et al (2000) Correlative light-electron microscopy reveals the tubular-saccular ultrastructure of carriers operating between Golgi apparatus and plasma membrane. J Cell Biol 148:45–58. https://doi.org/10.1083/jcb.148.1.45
7. Mironov AA, Polishchuk RS, Luini A (2000) Visualizing membrane traffic in vivo by combined video fluorescence and 3D electron microscopy. Trends Cell Biol 10:349–353. https://doi.org/10.1016/s0962-8924(00)01787-6
8. van den Dries K, Fransen J, Cambi A (2022) Fluorescence CLEM in biology: historic developments and current super-resolution applications. FEBS Lett 596:2486–2496. https://doi.org/10.1002/1873-3468.14421
9. Olmos Y, Hodgson L, Mantell J et al (2015) ESCRT-III controls nuclear envelope reformation. Nature 522:236–239. https://doi.org/10.1038/nature14503
10. Parton RG (1995) Rapid processing of filter-grown cells for Epon embedding. J Histochem Cytochem 43:731–733. https://doi.org/10.1177/43.7.7608529
11. Shewring JR, Hodgson L, Bryant HL et al (2021) Refining a correlative light electron microscopy workflow using luminescent metal complexes. Methods Cell Biol 162:69–87. https://doi.org/10.1016/bs.mcb.2020.12.008
12. Paul-Gilloteaux P, Heiligenstein X, Belle M et al (2017) eC-CLEM: flexible multidimensional registration software for correlative microscopies. Nat Methods 14:102–103. https://doi.org/10.1038/nmeth.4170
13. Krentzel D, Elphick M, Domart M-C et al (2023) CLEM-Reg: an automated point cloud based registration algorithm for correlative light and volume electron microscopy. bioRxiv. https://doi.org/10.1101/2023.05.11.540445
14. Pope I, Tanner H, Masia F et al (2023) Correlative light-electron microscopy using small gold nanoparticles as single probes. Light Sci Appl 12:80. https://doi.org/10.1038/s41377-023-01115-4
15. Hodgson L, Nam D, Mantell J et al (2014) Retracing in correlative light electron microscopy: where is my object of interest? Methods Cell Biol 124:1–21. https://doi.org/10.1016/B978-0-12-801075-4.00001-X
16. Müller-Reichert T, Verkade P (2012) Introduction to correlative light and electron microscopy. Methods Cell Biol 111:xvii–xix. https://doi.org/10.1016/B978-0-12-416026-2.03001-6
17. Müller-Reichert T, Verkade P (2014) Preface. Correlative light and electron microscopy II. Methods Cell Biol 124:xvii–xviii. https://doi.org/10.1016/B978-0-12-801075-4.09983-3
18. Müller-Reichert T, Verkade P (2017) Preface. Correlative light and electron microscopy III. Methods Cell Biol 140. https://doi.org/10.1016/bs.mcb.2017.03.007
19. Müller-Reichert T, Verkade P (2021) Preface to CLEM IV: broaden the horizon. Methods Cell Biol 162:xix. https://doi.org/10.1016/S0091-679X(21)00029-7
20. Ando T, Bhamidimarri SP, Brending N et al (2018) The 2018 correlative microscopy techniques roadmap. J Phys D Appl Phys 51:443001. https://doi.org/10.1088/1361-6463/aad055

21. Correlative Imaging, focussing on the future, Edited by Paul Verkade and Lucy Collinson (2019). ISBN: 978-1-119-08645-1
22. Brown E, Mantell J, Carter D et al (2009) Studying intracellular transport using high-pressure freezing and correlative light electron microscopy. Semin Cell Dev Biol 20:910–919. https://doi.org/10.1016/j.semcdb.2009.07.006
23. Brown E, Van Weering J, Sharp T et al (2012) Capturing endocytic segregation events with HPF-CLEM. Methods Cell Biol 111:175–201. https://doi.org/10.1016/B978-0-12-416026-2.00010-8

Chapter 9

In Situ Imaging of Proteins Using DNA-PAINT Super-Resolution Microscopy

Florian Schueder and Ralf Jungmann

Abstract

The spatial resolution of conventional light microscopy is restricted by the diffraction limit to hundreds of nanometers. Super-resolution microscopy enables single digit nanometer resolution by circumventing the diffraction limit of conventional light microscopy. DNA point accumulation for imaging in nanoscale topography (DNA-PAINT) belongs to the family of single-molecule localization super-resolution approaches. Unique features of DNA-PAINT are that it allows for sub-nanometer resolution, spectrally unlimited multiplexing, proximity detection, and quantitative counting of target molecules. Here, we describe prerequisites for efficient DNA-PAINT microscopy.

Key words Fluorescence microscopy, Super-resolution microscopy, DNA-PAINT, Microtubule imaging, Mitochondria imaging, Single-molecule localization microscopy, How-to DNA-PAINT

1 Introduction

Super-resolution microscopy enables single digit nanometer resolution by circumventing the diffraction limit of conventional light microscopy. It has been almost two decades since the invention of single-molecule localization microscopy (SMLM) [1, 2], which has fueled many exciting technical developments, pushing the limits of fluorescence microscopy in many ways. DNA point accumulation for imaging in nanoscale topography (DNA-PAINT) belongs to the family of single-molecule localization super-resolution approaches [3]. Unique features of DNA-PAINT are that it allows for sub-nanometer resolution [4, 5], spectrally unlimited multiplexing [6], proximity detection [7], and quantitative counting [8] of target molecules. There has been good effort from the SMLM community to reduce obstacles for novice users in super-resolution microscopy such as commercialization of reagents and microscopes, easy-to-use open-source software for data processing or detailed guides for different parts of how to perform

Christoph Wülfing and Robert F. Murphy (eds.), *Imaging Cell Signaling*, Methods in Molecular Biology, vol. 2800, https://doi.org/10.1007/978-1-0716-3834-7_9,

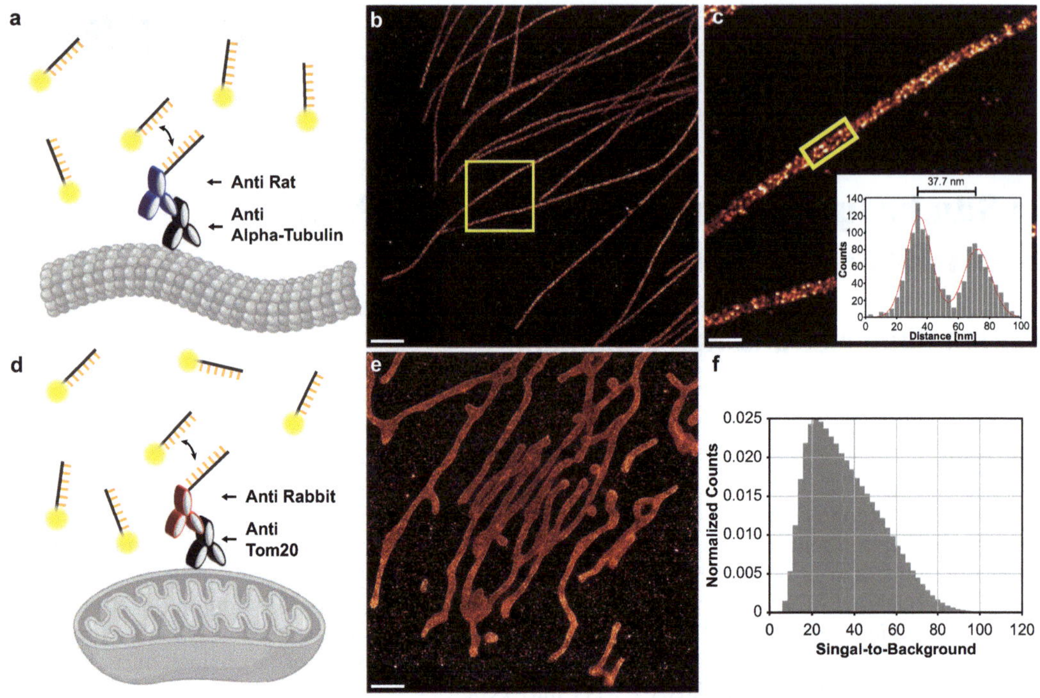

Fig. 1 DNA-PAINT imaging examples. (**a**) Schematic depiction of labeling of microtubules via primary (anti-alpha-tubulin) and secondary antibodies for DNA-PAINT imaging. (**b**) Exemplary field of view of tubulin filaments in a COS-7 cell acquired with DNA-PAINT. (**c**) Zoom in of the highlighted area in (**b**). The insert presents a histogram projection of the highlighted area and reveals the well-known "hollowness" of the microtubules. (**d**) Labeling concept for imaging the outer membrane protein Tom20 of mitochondria via DNA-PAINT. (**e**) Super-resolved mitochondria image in a HeLa cell (target protein Tom20). (**f**) Signal-to-background histogram of Tom20 data presented in (**e**). Scale bars, 1 μm (**b**) and (**e**), 200 nm (**c**)

super-resolution imaging. The intention of this article is to enable a person with minimal super-resolution imaging experience to acquire DNA-PAINT super-resolution images of organelles or filaments. In DNA-PAINT, transient binding of short dye-labeled ("imager") oligonucleotides to their complementary target ("docking") strands, commonly conjugated to an antibody, creates the necessary "blinking" to enable stochastic super-resolution microscopy. In this article, we discuss how to get started with a DNA-PAINT imaging experiment by splitting the process into sample preparation, imaging acquisition, and reconstruction as well as post processing. We explain every step using the imaging of microtubules and mitochondria in adherent cells as an example (Fig. 1).

2 Materials

2.1 DNA-PAINT Sequences for Imagers and Docking Sites

Currently, there are three different types of imager designs. The classical DNA-PAINT imagers in which more than 50 orthogonal sequences have been published [9]. Six imagers optimized for speed with an up to 100-fold higher association rate [10, 11] compared to the classical imagers. Two fluorogenic imagers increase the signal-to-background ratio by reducing the fluorescence of an unbound imager strand compared to a target-bound one up to 57-fold [12]. In the following table (Table 1), we give six examples for DNA-PAINT imagers and their corresponding docking sites. The name of the two imagers used in the imaging example in this chapter are P3 and PS3.

2.2 Solutions, Reagents, and Consumables

1. Imaging Buffer: 1× PBS + 500 mM NaCl. Store at RT.
2. 10% Triton-X 100 in 1× PBS. Store at 4 °C.
3. 5% BSA in 1× PBS. Store at 4 °C.
4. 100 kDa Amicon filter (Sigma, UFC510008)
5. DBCO-sulfo-NHS crosslinker (Jena Bioscience, CLK-A124-10)
6. Zeba column (Thermo Fisher Scientific, 89882)
7. 8-Well chambered cover glasses (ibidi, 80807)
8. Organelle fixative: 3% Paraformaldehyde (PFA) (Electron Microscopy Sciences, 15710) + 0.1% Glutaraldehyde (GA) (Electron Microscopy Sciences, 16219) in 1× PBS solution, prepare fresh

Table 1
Examples of DNA-PAINT imagers. The sequences and fluorescent dyes of six imagers are given, along with their name, type, and corresponding docking sites

Name	Type	Imager	Docking site (AB = Antibody)
P1	Classical	5-TAGATGTAT-Cy3B-3	5-AB-TT ATACATCTA-3
P3	Classical	5-TAATGAAGA-Cy3B-3	5-AB-TT TCTTCATTA-3
P5	Classical	5-ATACATTGA-Cy3B-3	5-AB-TT TCAATGTAT-3
PS3	Speed	5-TCCTCCC-Cy3B-3	5-AB-AA GGGAGGA-3
R2	Speed	5-TGGTGGT-Cy3B-3	5-AB-AA ACCACCACCACCACCACCA-3
FP2	Fluorogenic	5-Cy3B-AAGAAGTAAAGGGAG-BHQ2-3	5-CCTCGCTGAACCCCTTA TT-AB-3

9. Cytoskeleton fixative: 0.4% Glutaraldehyde (GA) (Electron Microscopy Sciences, 16219) + 0.25% Triton-X 100 in in 1× PBS, prepare fresh
10. 3% Glutaraldehyde (Electron Microscopy Sciences, 16219) in 1× PBS, prepare fresh
11. Blocking solution: 3% BSA + 0.25% Triton-X 100 in 1× PBS
12. Anti-Tom20 antibody (Abcam, ab78547)
13. Anti-alpha tubulin antibody (Sigma, T5168)
14. Gold particles with a diameter of 90 nm (Nanopartz, A11-90)

2.3 Microscope

In general, every microscope setup capable of "total internal reflection fluorescence" (TIRF) is suitable for DNA-PAINT imaging. In the following, we list a couple of hardware features, which are not strictly necessary but enable high-performance DNA-PAINT imaging.

1. TIRF microscope with a TIRF objective lens
2. sCMOS or EMCCD camera
3. Laser >200 mW (nominal) 561 nm and 641 nm
4. Active Focus Lock (e.g., Perfect Focus System from Nikon)

A more detailed description of a suitable setup for DNA-PAINT is described in Schnitzbauer et al. [13].

3 Methods

3.1 Secondary Antibody Conjugation

When imaging proteins, secondary antibodies need to be conjugated to a single-stranded DNA oligo, which functions as a docking site for the complimentary DNA-PAINT imager.

1. Add 500 μL of 1× PBS to a 100 kDa Amicon filter
2. Let filter equilibrate for 1 h at room temperature (RT).
3. Spin down 1× PBS at 14,000 g for 5 min and discard the flow through.
4. Add 500 μL of secondary antibody (e.g., Anti-Rabbit, c = 1 mg/mL) to 100 kDa Amicon filter and spin down at 14,000 g for 5 min to concentrate.
5. Extract the residual solution in the Amicon filter by placing the Amicon filter upside down into a new tube and spin at 1000 g for 2 min.
6. Adjust the volume to 100 μL by adding 1× PBS.
7. Check concentration via absorption using, for example, a UV/VIS absorption spectrometer. The concentration is expected to be >2.5 mg/mL (*see* **Note 1**).

8. Mix concentrated antibodies with DBCO-sulfo-NHS cross-linker in a 1:10 ratio (*see* **Note 2**).
9. Age this reaction for ~1.5 h at 4 °C on a rotator. Keep away from light.
10. After 1.5 h, remove the excess of cross-linker using a Zeba column at 4 °C.
11. Spin down the Zeba column at room temperature at 1500 g for 1 min to remove the storage buffer.
12. Wash Zeba column three times with 300 μL of 1× PBS by spinning down at 1500 g for 1 min.
13. Remove excess cross-linker from antibody by spinning it down in the Zeba column at 1500 g for 2 min (*see* **Note 3**).
14. Check Antibody concentration via absorption.
15. Add Azide-modified DNA (at the 5-prime end) to the modified antibody in a ratio 1:10 molar ratio (*see* **Note 4**). Age reaction overnight at 4 °C on a rotator. Keep away from light.
16. On the next day, add 500 μL of 1× PBS to a 100 kDa Amicon filter. Let filter equilibrate for 1 h at room temperature.
17. Purify DNA-labeled antibody (AB) from DNA excess via a 100 kDa Amicon filter. Spin down 1× PBS at 14,000 g for 5 min and discard flow through four times.
18. Dilute antibody DNA construct with 100% glycerol (final glycerol concentration >50%).
19. Store conjugated antibody at −20 °C. At −20 °C, the antibodies can be stored at least for 6 months.

A more detailed description can be found in Agasti et al. [9] or Schnitzbauer et al. [13]. Alternatively, ready-to-use DNA-conjugated antibodies are commercially available from Massive Photonics (www.massive-photonics.com).

3.2 Sample Preparation

1. Seed ~30,000 cells (*see* **Note 5**) in 8-well chambered cover glasses (*see* **Note 6**). Grow them overnight at 37 °C and 5% CO_2 in an incubator.
2. After ~24 h, the cells are ready for fixation.
3. Fixation is a critical step in sample preparation (*see* **Note 7**). Execute either **step 4** or **5**.
4. Organelle fixation: Fix with organelle fixative (prewarmed to 37 °C) for 15 min rocking at RT.
5. Cytoskeleton fixation. Pre-fix and extract with cytoskeleton fixative (prewarmed to 37 °C) for ~90 s at RT. Replace the pre-fix buffer with 3% GA solution and incubate for 15 min on a rocker at RT.

6. Wash samples four times with 1× PBS (30 s, 60 s, 2× 5 min) on a rocker at RT.
7. Block samples with blocking solution for 1 h at RT.
8. Incubate with primary antibody Tom20 or alpha-tubulin (1: 200 dilution) in blocking solution on a rocker at 4 °C overnight.
9. On the next day, wash samples four times with 1× PBS (30 s, 60 s, 2× 5 min) on a rocker at RT.
10. Incubate with DNA-labeled secondary antibodies (1:200 dilution) corresponding to the primary antibody used in **step 6** in blocking solution on a rocker at RT for 2 h (*see* **Note 8**).
11. Wash samples four times with 1× PBS (30 s, 60 s, 2× 5 min) on a rocker at RT.
12. Incubate samples for 5 mins with gold particles with a diameter of 90 nm (dilution 1:100) in 1× PBS. These gold particles can later be used as fiducial markers for drift correction. Sonicate gold particles for 5 min before diluting them in 1× PBS.
13. Rinse sample four times with 1× PBS.
14. Replace 1× PBS with imaging buffer solution containing the DNA-PAINT imager. The concentration of the imager depends on the sequence being used and on the target protein density (*see* **Note 9**). Execute either **step 15** or **16**.
15. For the microtubule (alpha-tubulin) example, we used the imager sequence PS3 labeled with a Cy3B at a concentration of 500 pM in imaging buffer.
16. For the mitochondria (Tom20) example, we used the imager sequence P3 labeled with a Cy3B at a concentration of 2 nM.

3.3 Imaging

1. To find a field-of-view (FOV) of interest, microtubules can typically be imaged in a plane close to the coverslip. Mitochondria can be imaged in a range ~500 nm to 1 μm away from the coverslip.
2. Adjusting the illumination. The further the plane of interest is away from the coverslip the more the TIRF angle needs to be tuned from the critical angle of "total-internal-reflection-fluorescence" (TIRF) to "highly-inclined-and-laminated-optical' (HILO) [14] sheet. As microtubules are typically imaged best in a region close to the coverslip, they can be imaged mostly in TIRF illumination. Imaging of mitochondria works best in HILO illumination.
3. Setting the excitation intensity. The excitation intensity is important as the number of photons of the localization events determines the downstream resolution of the super-resolved image. We suggest slowly increasing the laser intensity while

looking at the FOV in live imaging mode until the number of observed localization events per frame decreases (bleaching of imagers). From here, the intensity should be decreased by 5–10%. For the microtubule image (Fig. 1b), the excitation intensity was ~2 kW/cm^2. The excitation intensity for the mitochondria image was ~1.5 kW/cm^2.

4. Choosing an image acquisition duration. The total imaging time should be set to a duration such that every target molecule is imaged at least one time (usually tens of times) with a probability of >95% (*see* **Note 10**). For the sample microtubule image, the imaging time was 10 min. For the sample mitochondria image, the imaging time was 50 min.
5. Setting the camera frame rate. The camera frame rate depends on the binding time of the selected imager. In general, the exposure time, should not be much shorter than the average bound time of an imager probe. In the sample microtubule image, we imaged with a speed sequence (PS3); the exposure time was set to 20 ms. For imaging mitochondria with a classical DNA-PAINT sequence (P3), the exposure time was set to 100 ms.

3.4 Image Reconstruction and Postprocessing

For the image reconstruction, we suggest using Picasso [13] as it was specifically developed for DNA-PAINT imaging. However, any other software for single-molecule-based super-resolution reconstruction can be used [15]. In the following workflow, we describe how to reconstruct and postprocess super-resolution data with Picasso. This workflow is illustrated in Fig. 2.

1. If the data is in an OME-TIF (e.g., acquired with Micromanager [16]), the data can be opened via drag-and-drop in Picasso Localize (*see* **Note 11**).
2. Insert all camera parameters in the "Photon conversion" section in the parameters window.
3. Set the "Box side length" and "Min. Net. Gradient" that most localization events are detected, but no unspecific events are captured (e.g., background).
4. Choose as a fit method either "least square" or "maximum likelihood estimation." The first is usually recommended.
5. Localize (Identify & Fit) the data (Analyze → Localize).
6. The generated hdf5-file can be visualized with Picasso Render.
7. Picasso allows to correct for sample drift in two ways.
   (a) Drift correction by redundant cross-correlation (RCC). For the segmentation parameter, 10% of the total image duration is a good starting point (e.g., 1000 frames of a 10,000 frames data set) (Postprocess → Undrift by RCC).

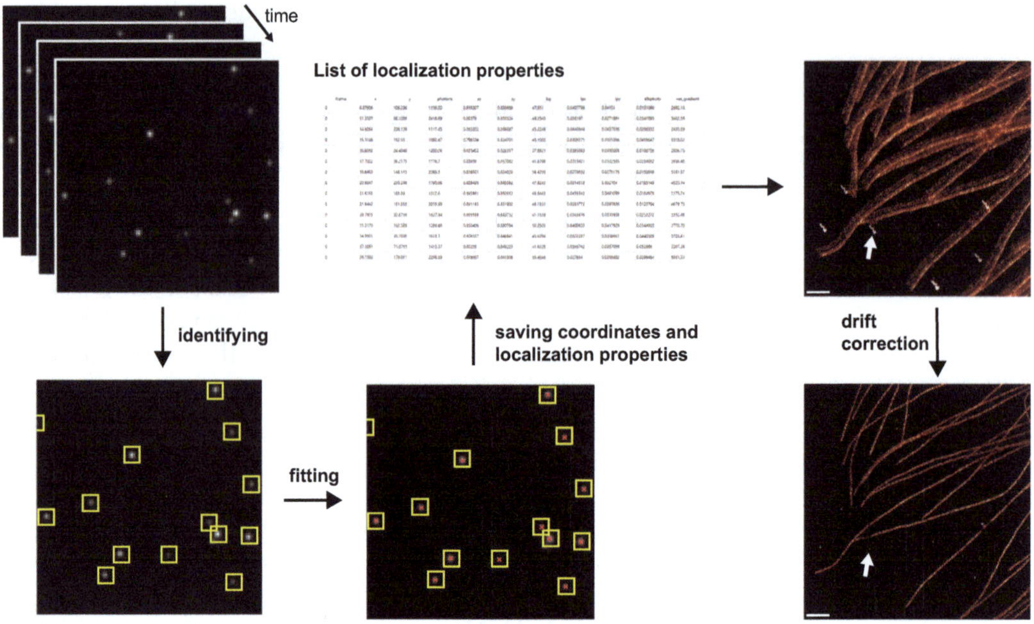

Fig. 2 Image reconstruction and postprocessing workflow. After acquiring a time series on the microscope, the raw data is first processed with Picasso Localize. First, all single-molecule events are identified, and in a second step, the individual events are fitted. The coordinates and other localization properties are saved in a list. This list of coordinates can then be visualized via Picasso Render. The data can be further postprocessed in Picasso Render, for example, correcting for sample drift. Scale bars, 1 μm

(b) Drift correction by fiducial marker (e.g., gold particles). The fiducial markers can be selected manually or semi-automatically using the pick tool. Once the fiducial markers are selected, the selection can be used for drift correction by selecting "Undrift from pick" (Postprocess → Undrift from pick).

8. The localization precision and downstream the resolution of the super-resolved image can be estimated via the NeNA [17] metric (View → Show Info →Nena precision).
9. The super-resolved image can be exported as a PNG-file via Picasso Render (File → Export complete image).

4 Notes

1. Measuring antibody concentration via absorption can be done using for example a Nanodrop absorption spectrometer.
2. Before opening the cross-linker, it is important to check the instructions on how to resuspend and how to store the cross-linker. We suggest preparing a crosslinker stock solution at 10 mg/mL resuspended in DMF and aliquot the solution in

30-μL fractions. The solution can be stored at −80 °C for 3 months. Make sure not to keep them at RT for more than 15 min as the cross-linker will lose its ability to react.

3. Before using the Zeba column to purify the antibody from the excess cross-linker, we recommend checking the instructions. The Zeba column used in the methods part can only hold a volume of 130 μL. If the volume of the antibody crosslinker mix is more than 120 μL, we recommend using multiple Zeba columns for purification.
4. Calculate the molar ratio between antibody and DNA.
   (a) The concentration of the antibody after the concentration step is ~4 mg/mL. The molecular weight of a secondary antibody is ~150 kDa = $150*10^3$ g/mol. The molar concentration of the antibody is therefore $c = \frac{4\frac{g}{L}}{150*10^3\frac{g}{\mathrm{mol}}} = 26.67\ \mu\mathrm{M}$.
   (b) The DNA concentration can be adjusted. We recommend resuspending the lyophilized DNA to a concentration of 1 mM.
   (c) To achieve a 1:10 molecular ratio between antibody to DNA, mix 100 μL of antibody at a concentration of 26.67 μM with 26.7 μL of DNA at 1 mM.
5. In general, DNA-PAINT imaging works well in adherent cells, for example, HeLa cells, U-2 OS cells, BS-C-1 cells, or COS-7 cells.
6. For a high-quality DNA-PAINT super-resolution experiment, it is important to use high precision glass coverslips (i.e., no polymer coverslips) with a 170-μm thickness, for example, #1.5.
7. Important for optimal fixation is to always prepare the fixation buffer fresh. We suggest opening a vial of PFA or/and GA right before the fixation step. Moreover, the fixation solution needs to be optimized to the target protein of interest. In this protocol, we describe a fixation strategy for membrane proteins of organelles like Tom20 (Mitochondria) and a second fixation strategy for cytoskeletal proteins, for example, tubulin or vimentin.
8. Primary antibody made against a specific protein of interest are typically obtained from a host animal (e.g., mouse, rabbit, rat, chicken, goat). The secondary antibody binds against all primary antibodies from a specific host. For example, the Tom20 antibody was raised in rabbit as host, which means the corresponding secondary antibody would be an anti-rabbit antibody.

9. Selecting a DNA-PAINT imager concentration mainly depends on the target protein density and the used imager DNA sequence. The higher the density of a target protein is, the lower the concentration of the imager must be to avoid artifacts from multiple localization events (two or more events in a diffraction limited volume at the same time). In practice, this is an empirical value and must be determined for every target protein individually. The imager concentration depends further on the association and the dissociation rate of the imager to a corresponding docking site. If, for example, the association rate is increased by using a speed DNA-PAINT imager, this must be compensated by increasing the dissociation rate or decreasing the concentration of the imager in order to avoid double or multiple binding events in a diffraction-limited volume.
10. Calculating the duration to achieve a sampling probability >95%.
    (a) The sampling probability P can be calculated $P = 1 - e^{-\frac{t}{\tau_D}}$ where τ_D is the average time in between binding events and t is the image acquisition duration.
    (b) We suggest that $t \approx 3\tau_D$.
    (c) For an association rate of PS3, $k_{\text{on}} = 1 * 10^7 \frac{1}{M\ \text{sec}}$, for example, the average time between binding events for a concentration of $c = 500$ pM is $\tau_D \approx 200$ sec. This suggests that for a sampling probability of at least 95% the time of image acquisition must be at least 600 s or 10 min.
11. If the microscope saves the imaging data in another format as OME-TIF, Fiji [18] can be used to open the data, convert, and save it as a RAW-File. This RAW-File can then be opened and processed with Picasso Localize.

Acknowledgments

F.S. acknowledges support from the Human Frontier Science Program (LT000056/2020-C). R.J. acknowledges support from the Max Planck Society.

References

1. Betzig E, Patterson GH, Sougrat R et al (2006) Imaging intracellular fluorescent proteins at nanometer resolution. Science 313:1642–1645. https://doi.org/10.1126/science.1127344
2. Rust MJ, Bates M, Zhuang X (2006) Sub-diffraction-limit imaging by stochastic optical reconstruction microscopy (STORM). Nat Methods 3:793–795. https://doi.org/10.1038/nmeth929
3. Lelek M, Gyparaki MT, Beliu G et al (2021) Single-molecule localization microscopy. Nat Rev Methods Primers 1. https://doi.org/10.1038/s43586-021-00038-x

4. Dai M, Jungmann R, Yin P (2016) Optical imaging of individual biomolecules in densely packed clusters. Nat Nanotechnol 11:798–807. https://doi.org/10.1038/nnano.2016.95
5. Reinhardt SCM, Masullo LA, Baudrexel I et al (2023) Angstrom-resolution fluorescence microscopy. Nature 617:711–716. https://doi.org/10.1038/s41586-023-05925-9
6. Jungmann R, Avendano MS, Woehrstein JB et al (2014) Multiplexed 3D cellular super-resolution imaging with DNA-PAINT and exchange-PAINT. Nat Methods 11:313–318. https://doi.org/10.1038/nmeth.2835
7. Schueder F, Lara-Gutierrez J, Haas D et al (2021) Super-resolution spatial proximity detection with proximity-PAINT. Angew Chem Int Ed Engl 60:716–720. https://doi.org/10.1002/anie.202009031
8. Jungmann R, Avendano MS, Dai M et al (2016) Quantitative super-resolution imaging with qPAINT. Nat Methods 13:439–442. https://doi.org/10.1038/nmeth.3804
9. Agasti SS, Wang Y, Schueder F et al (2017) DNA-barcoded labeling probes for highly multiplexed exchange-PAINT imaging. Chem Sci 8:3080–3091. https://doi.org/10.1039/c6sc05420j
10. Schueder F, Stein J, Stehr F et al (2019) An order of magnitude faster DNA-PAINT imaging by optimized sequence design and buffer conditions. Nat Methods 16:1101–1104. https://doi.org/10.1038/s41592-019-0584-7
11. Strauss S, Jungmann R (2020) Up to 100-fold speed-up and multiplexing in optimized DNA-PAINT. Nat Methods 17:789–791. https://doi.org/10.1038/s41592-020-0869-x
12. Chung KKH, Zhang Z, Kidd P et al (2022) Fluorogenic DNA-PAINT for faster, low-background super-resolution imaging. Nat Methods 19:554–559. https://doi.org/10.1038/s41592-022-01464-9
13. Schnitzbauer J, Strauss MT, Schlichthaerle T et al (2017) Super-resolution microscopy with DNA-PAINT. Nat Protoc 12:1198–1228. https://doi.org/10.1038/nprot.2017.024
14. Tokunaga M, Imamoto N, Sakata-Sogawa K (2008) Highly inclined thin illumination enables clear single-molecule imaging in cells. Nat Methods 5:159–161. https://doi.org/10.1038/nmeth1171
15. Sage D, Pham TA, Babcock H et al (2019) Super-resolution fight club: assessment of 2D and 3D single-molecule localization microscopy software. Nat Methods 16:387–395. https://doi.org/10.1038/s41592-019-0364-4
16. Edelstein A, Amodaj N, Hoover K et al (2010) Computer control of microscopes using micro-Manager. Curr Protoc Mol Biol Chapter 14: Unit14 20. https://doi.org/10.1002/0471142727.mb1420s92
17. Endesfelder U, Malkusch S, Fricke F, Heilemann M (2014) A simple method to estimate the average localization precision of a single-molecule localization microscopy experiment. Histochem Cell Biol 141:629–638. https://doi.org/10.1007/s00418-014-1192-3
18. Schindelin J, Arganda-Carreras I, Frise E et al (2012) Fiji: an open-source platform for biological-image analysis. Nat Methods 9: 676–682. https://doi.org/10.1038/nmeth.2019

Chapter 10

A Magnetic Pincher for the Dynamic Measurement of the Actin Cortex Thickness in Live Cells

Joseph Vermeil, Valentin Laplaud, Anumita Jawahar, Dulamkhuu Bujaa, Damien Cuvelier, Julien Heuvingh, Olivia du Roure, and Matthieu Piel

Abstract

The actin cortex is an essential element of the cytoskeleton allowing cells to control and modify their shape. It is involved in cell division and migration. However, probing precisely the physical properties of the actin cortex has proved to be challenging: it is a thin and dynamic material, and its location in the cell—directly under the plasma membrane—makes it difficult to study with standard light microscopy and cell mechanics techniques. In this chapter, we present a novel protocol to probe dynamically the thickness of the cortex and its fluctuations using superparamagnetic microbeads in a uniform magnetic field. A bead ingested by the cell and another outside the cell attract each other due to dipolar forces. By tracking their position with nanometer precision, one can measure the thickness of the cortex pinched between two beads and monitor its evolution in time. We first present the set of elements necessary to realize this protocol: a magnetic field generator adapted to a specific imaging setup and the aforementioned superparamagnetic microbeads. Then we detail the different steps of a protocol that can be used on diverse cell types, adherent or not.

Key words Actin cortex, Cell mechanics, Magnetic beads, Image analysis, Dynamic measurement

1 Introduction

The actin cortex is a thin and dense meshwork of actin filaments located just underneath the plasma membrane of eukaryotic cells. It is a key element of the cytoskeleton in many processes that involve a control of the cell shape. Indeed, the cortex is an active material, due to the actin filament fast turnover and their association with myosin II microfilament. The gradients of tension that these motors generate are the source of cell-scale deformations that occur, for instance, during bleb-mediated amoeboid migration or furrowing during cell division [1]. Because the cortex is very thin relatively to the cell and the tensile forces it generates are mainly parallel to the cell surface, the cortex is often viewed as a two-dimensional sheet of actin. However, this point of view

Christoph Wülfing and Robert F. Murphy (eds.), *Imaging Cell Signaling*, Methods in Molecular Biology, vol. 2800, https://doi.org/10.1007/978-1-0716-3834-7_10,

neglects the role of the 3D architecture of the cortex, which is key for many of its emerging properties. This architecture is very challenging to study experimentally: single-actin filaments are below the diffraction limit, their density in the cortex is high, and their typical turnover rate is a few tens of seconds [2]. As a whole, the cortex is very dense in actin filaments and contains a host of other proteins that associate with actin to produce different types of structures. This explains why the most precise observations of the cortex have been obtained with electron microscopy on fixed samples, cells (spread cells [3], round cells [4]), or cellular blebs [5]. More recently, techniques compatible with live cells were developed: atomic force microscopy [6] or sub-resolution microscopy [4, 7] helped to better understand the regulation mechanisms of cortical actin architecture, by combining observations with perturbations of actin-binding proteins activity using drugs or siRNA.

We propose here a new approach, relying on magnetic microbeads to pinch the cortex of live cells. These beads, commonly used for cell separation, acquire a magnetization when placed in an external magnetic field. If a bead is inside a cell and another outside, they will form a pair pinching the cortex by attracting each other. When the pinching force is kept low and constant, the beads do not significantly indent the cortex. In this configuration, by tracking their relative position and subtracting the size of the beads, one can probe the thickness of the cortex at the beads location with a very high spatial (10 nm) and temporal (sub-second) resolution. In particular, this allows us to monitor the fluctuations of the thickness in time, which are significant, and correlated with the activity of Myosin II motors in the actin meshwork [8]. Thanks to its versatility, this technique can be applied to many cell types in diverse contexts and setup. By basing this thickness measurement on a force application, our approach also open new ways to directly measure cortex mechanics.

2 Materials

2.1 Bead Preparation

1. M-450 Epoxy beads (Dynal, Thermo Fisher, USA).
2. Culture Medium 3 T3 Fibroblasts: Dulbecco's Modified Eagle Medium with GlutaMAX supplement, fetal calf serum at 10% of final volume (S1810-500, Biowest, France), penicillin-streptomycin at 1% of final volume, that is, 100 U/mL.
3. Culture Medium Primary Mouse Dendritic Cells: Iscove's modified Dulbecco's medium, fetal calf serum at 10% of final volume, (S1810-500, Biowest, France), glutamine (20 mM), penicillin-streptomycin at 1% of final volume, that is, 100 U/ mL and 2-mercaptoethanol (50 μm) supplemented with

granulocyte-macrophage colony-stimulating factor (50 ng/mL) containing supernatant obtained from transfected J558 cells, as previously described [8, 9].

2.2 Experimental Chamber Preparation (for 3 T3)

1. Holed petri dish (35 mm) *or* petri dish (35 mm) and access to a laser cutter.
2. PLL(20)-g[3.5]-PEG(2) (SuSoS, Switzerland).
3. HEPES buffer: 10-mM HEPES in MilliQ water, pH 7.4.
4. Fibronectin (from bovine serum, F1141, Merck, Germany).
5. Sodium bicarbonate buffer: 100-mM sodium bicarbonate in MilliQ water, pH 8.3.
6. Quartz photomask (JD Photodata, UK).
7. Deep UV-source (UVO Cleaner: $\lambda = 254$ nm, $P = 7$ mW/m^2; Jelight, USA).
8. Nontoxic glue (silicon glue "SA 500," Zolux, France).

2.3 Cell Handling

1. Culture medium: identical as above.
2. Imaging medium: identical to culture media. When trying to combine these experiments with fluorescence imaging, replacing DMEM with FluoroBrite (A1896701, Thermo Fischer, USA) in the medium composition can improve the image quality.

2.4 Other Equipment

1. Gaussmeter (model GM08 with transverse probe, Hirst Magnetic Instruments Ltd., UK).

3 Methods

3.1 Overview

The setup consists of an inverted microscope used in brightfield illumination, with a system mounted on the microscope stage to generate a uniform magnetic field at the location of the imaged sample. This system can be an array of permanent magnets or a pair of electromagnetic coils. The protocol, whose main steps are summarized on Fig. 1, is based on 4.5-μm superparamagnetic microbeads, the M-450 Dynabeads. Cells are incubated with these beads until a significant uptake is reached. Then, cells are transferred in the experimental chamber, together with additional beads. The chamber, placed on the microscope, is exposed to a controlled uniform magnetic field. The beads magnetize and attract each other at short range, forming pairs or chains of beads. This self-organization generates simultaneously a large number of pinching events in the chamber (Fig. 2a, d), meaning a situation where a bead inside the cell forms a pair with another outside the cell, which pinches the cell cortex. Such beads can then be filmed under

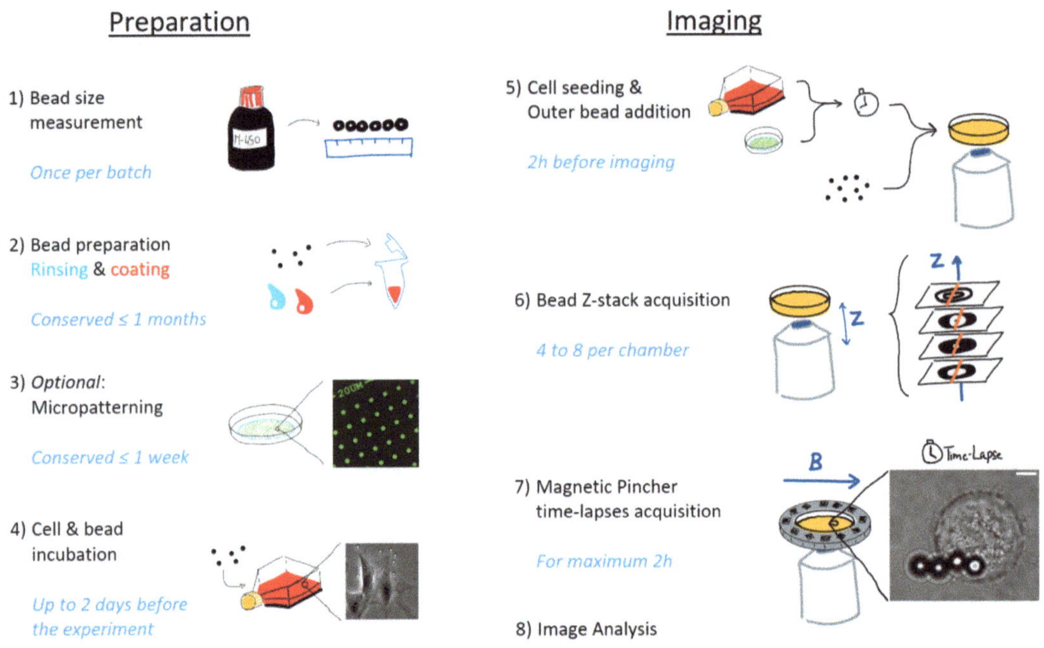

Fig. 1 Overview of the Magnetic Pincher steps. Schematics presenting the different steps of the method

brightfield illumination, and the images analyzed with a simple algorithm tracking the beads positions (Fig. 2b). This way we compute the cortical thickness and its evolution in time (Fig. 2c) with a high spatial ($\approx$ 30 nm) and temporal resolution (up to 100 Hz; more details in Ref. [8]).

3.2 Magnetic Field Generation

1. This protocol requires the generation of a uniform magnetic field over the experimental chamber. When exposed to an external magnetic field, superparamagnetic beads will become magnetic dipoles, with a magnetic moment that is a function of the applied field. Because the gradient of the external field is negligible, there is no long-range magnetic force, and the beads do not drift. Instead, the dipolar force generated by each bead causes them to attract each other. This leads to a self-organization of the beads present in the experimental chamber, forming pairs or chains aligned with the magnetic field (Figs. 3d and 5a, b). We insist here on the importance of the field uniformity: a too strong gradient of magnetic field would cause superparamagnetic beads to drift toward the higher field regions.

We propose below two options to generate a controlled and uniform magnetic field on a microscope. The first is the Halbach array [10], which consists in a set of permanent magnets arranged in a specific geometry around the sample. The second relies on two copper coils placed on both sides of the sample, in a pseudo-Helmholtz configuration.

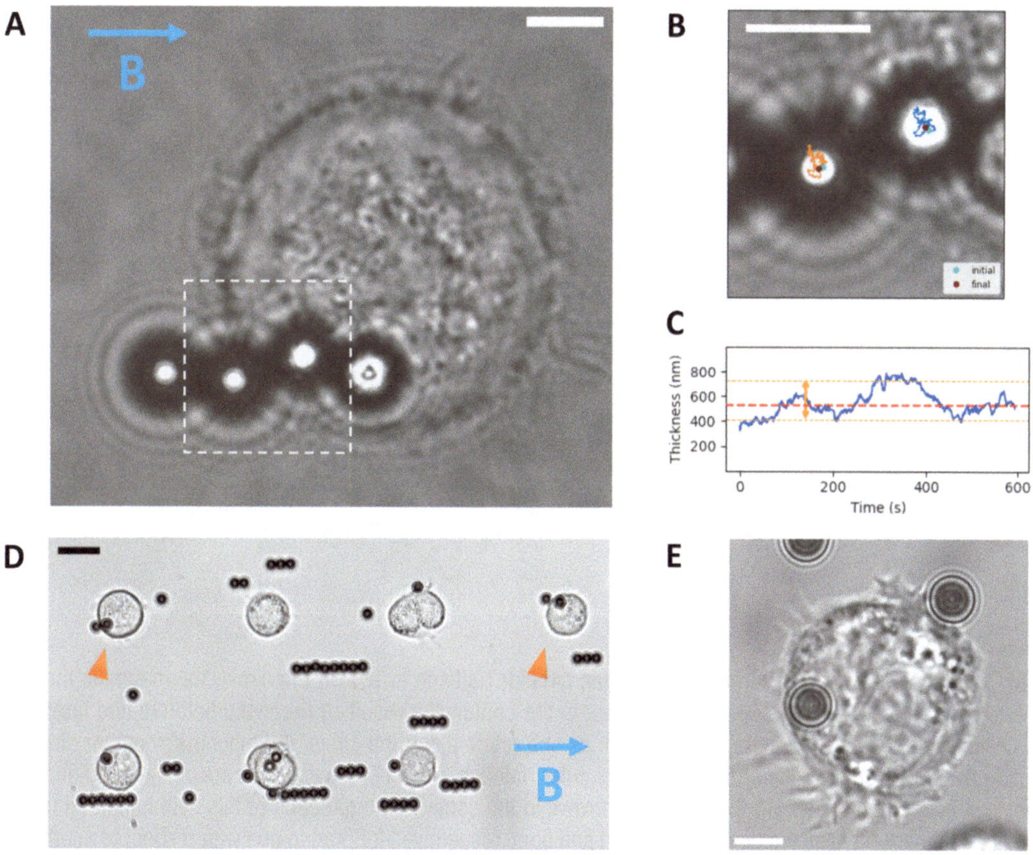

Fig. 2 Principle of the Magnetic Pincher technique. (**a**) 3 T3 fibroblast with its cortex pinched by super-paramagnetic M-450 Dynabeads. The image was acquired with brightfield microscopy, objective 100×, NA = 1.4, scale bar = 5 μm. The cell is adhering on a 20-μm micropatterned fibronectin disc and cannot spread more because the surrounding substrate is coated with PLL-g-PEG. The cell has ingested two beads, and two outer beads have been attracted toward them, the ensemble forming a chain aligned with the magnetic field (blue arrow). The two beads in the white frame are the ones pinching the cortex. Here, the focus is made on the light spot below the beads. NB: The bright spot is saturated in this image for better visualization, but should not be saturated during the course of the experiment. (**b**) A zoom on the region marked by the white frame, scale bar = 5 μm. The trajectories of the beads within a 10 min observation are shown in orange and blue. (**c**) Cortical thickness as a function of time (blue line) for this 3 T3 fibroblast. The median (red dashed line) and the inter-decile difference (D9 – D1, orange lines and arrow) are the metrics used to characterize respectively the typical cortical thickness and the fluctuation amplitude. Here, the median is 525 nm and the fluctuations amplitude 320 nm. (**d**) A larger field of view of a Magnetic Pincher experiment, with 3 T3 cells adhering on 20-μm fibronectin discs. The orange arrows point to two pinching event suitable for image acquisition. Magnification 40×, scale bar 20 μm. (**e**) A 3 T3 cell with one M-450 inside, with the focus made on the bead equatorial plane. The cell is adhering on a 20-μm fibronectin disc; scale bar 5 μm

2. Halbach Array

The term "Halbach array" refers to an arrangement of permanent multipole magnets. These arrays can have very diverse properties and application depending on their geometry. Here, we use a set of dipolar magnets arranged in a circle (for more details, see

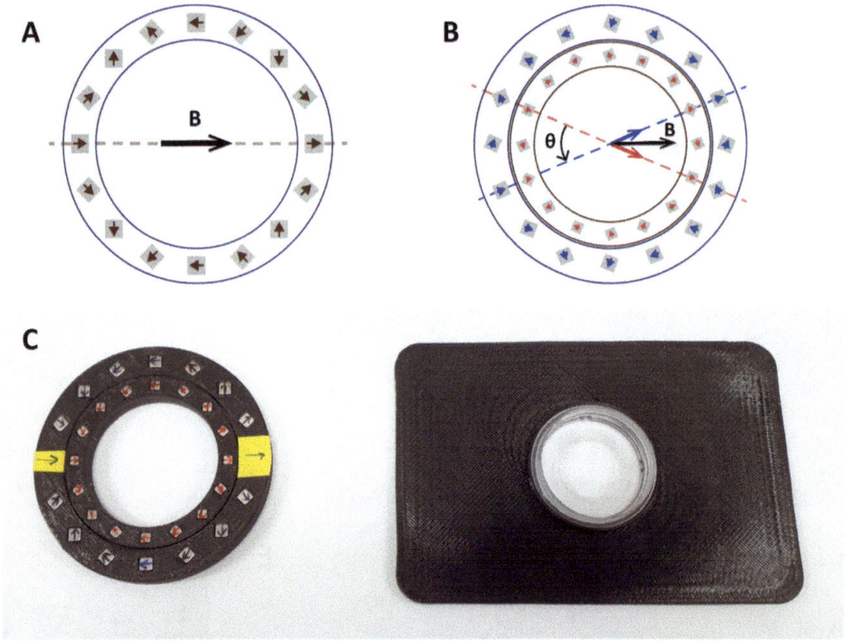

Fig. 3 Halbach array. (**a**) Schematic of a simple circular Halbach array. On the circle, the magnetic moment vectors of magnets are represented as arrows; in the center, the resulting magnetic field created inside the Halbach. Note the way magnets are arranged: as one rotate along the circle, the magnets' moment direction rotate twice as fast. (**b**) Schematic of nested circular Halbach arrays. The two arrays are designed to generate fields of equal magnitudes B0 (blue and red arrows in the center). The direction of this field is figured for the outer and inner arrays by the dashed lines (red and blue, respectively). Rotating one with respect to another by an angle θ allows to tune the magnitude of the total field (black arrow): $B = 2B0 \cdot \cos(\theta/2)$. (**c**) 3D-printed Halbach corresponding to Table 1 specifications. Right: the nested Halbach arrays. Left: The associated rectangular support that has the same outer dimensions as a 6-well plate. A central cylindrical extrusion allows to fit an experimental chamber inside (here a micropatterned chamber, *see* Subheading 1.3.5) and the Halbach arrays outside

[11], Fig. 3 and specifications in Table 1). It generates a uniform magnetic field inside the circle and a negligible field outside. The magnets have to be identical in size and magnetization and arranged so that their direction vector rotates twice faster than their position vector on the circle (Fig. 3a).

In such geometry, the field generated in the center of the Halbach array has the following magnitude:

$$B_0 = \frac{3}{8} N \mu_0 \frac{m}{\pi . R^3} \tag{1}$$

where N is the number of magnets used in the array, μ_0 is the vacuum magnetic permeability, R is the radius of the circle joining the center of magnets, and m is the magnitude of the magnetic moment of the magnets.

Another useful relationship in Halbach array design is the following: for a given magnet of volume V that contains a uniform

Table 1
Example of nested Halbach array design. Technical specifications

Support	
Length × width × thickness	127 × 85 × 3.5 mm
Inner array	
Inner/central/outer radius	21/24.75/28 mm
Height	5 mm
Magnet side length	3 mm
Number, type of magnets	16, N42 neodymium
Outer array	
Inner/central/outer radius	28/33/38 mm
Height	5 mm
Magnet side length	4 mm
Number/type of magnets	16/N42 neodymium
Magnetic properties	
Maximum magnetic field magnitude (arrays in the same direction)	Ideally: 8.6 mT Experimentally: 8.3 mT
Minimum magnetic field magnitude (arrays in opposite directions)	Ideally: 0 mT Experimentally: 0.2 mT
Gradient over the central 1 cm region	< 0.11 $mT.mm^{-1}$

field of magnetization *M*, the resulting magnetic moment m is simply:

$$m = V.M \tag{2}$$

As a reference, the neodymium N42, of which the cubic magnets we use are made, has a magnetization of $M_{N42} \approx 1.01 \times 10^6$ A/m. Their magnetization is simply deduced from this value and their geometry.

3. A slightly more complex application of this concept is the nested Halbach array (Fig. 3b). The idea is two use two concentric Halbach array that generate the same field B_0 and can be rotated independently. This is possible using larger magnets for the outer array, since, as stated by Eq. 1, the generated field magnitude decreases rapidly when the radius of the array increases; this decrease is compensated by increasing the volume of the magnets, so they possess a larger magnetic moment (Eq. 2). Therefore, this system can generate uniform magnetic field of any magnitude between ≈ 0 mT (when the two arrays'

fields are in opposite direction) and $2.B_0$ (when they are in the same direction).

We detail here the specifications of a nested array that we have optimized for this protocol and that is easy to mount on many microscopes, given that it has the outer dimension of a 6 well plate (Table 1, Fig. 3c). The body of the device, meaning the rectangular support and the two rings bearing the magnets, has been 3D-printed (Printer: Fortus 250mc, Stratasys, USA; Material: ABS X-TREME, iSQUARED, Switzerland). The neodymium cubic magnets are commercially available (Supermagnete, Germany). This instance of the nested Halbach array is designed to generate a maximum field of $2.B_0 = 8.6$ mT.

4. Coils

As a consequence of Ampere's law, an electric current circulating through a coil of conductive wires generate a magnetic field that curls around the coils. A pair of such coils positioned symmetrically along a common axis, with a distance between the coils of the same order of magnitude than the radius of the coils, are called pseudo-Helmoltz coils. If the same currents circulate through both coils in the same direction, a quasi-uniform magnetic field will be generated in the space between the coils. The magnetic field magnitude in the center of this system is proportional to the intensity of the electric current. Hence, such system can generate a field that can be tuned directly by controlling this current supplied to the coils.

In our case, the two coaxial coils (custom made by SBEA Technologies, France) are completed with a mu metal core (750 spires; length: 40 mm; inner diameter: 46 mm; outer diameter: 86 mm, see Fig. 4a) to increase the generated field. The coils are powered by a bipolar operational power supply amplifier 6A/36 V (Kepco, USA) controlled by the computer through a data acquisition module (National Instruments, USA). The maximum field generated is 55 mT (which correspond to the maximum supplied current, 6 A) with a gradient less than 0.1 $mT{\cdot}mm^{-1}$ over the sample.

5. Comparison of the two solutions (*see* also Table 2)

Although both options can produce a similar uniform magnetic field in the sample zone, they offer different advantages and disadvantages, which are summarized in the Table 2 below. The most crucial difference is that the Halbach array is easier to make and to use on any microscope, while the coils require special manufacturing and a custom-made microscope stage (Fig. 4b). On the other hand, the magnetic field generated by the coils can be dynamically adjusted during an experiment (opening the doors to rheology experiments, which are beyond the scope of the method described here), which is impossible with a Halbach array.

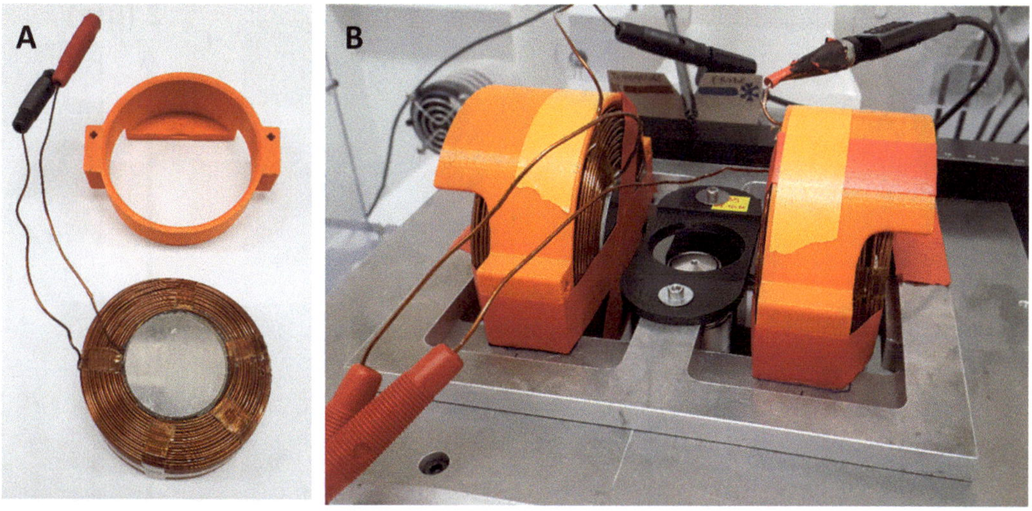

Fig. 4 Coils and imaging setup. (**a**) A coil and its case. The coil (bottom) itself is comprised of the mu-metal core, the spires, and the electric connectors. The role of the case (top) is to support the coil when it is mounted on the microscope stage. (**b**) A pair of coils mounted on an inverted microscope stage. The coils in their cases are coaxial and equidistant from the objective. The stage comprises a rectangular hole to allow the coils to be mounted. Its mobile part is attached to an XY micromanipulator and support a 3D-printed dish-holder

Table 2
Comparison of the two magnetic field generation solution

	Halbach array	Coils
Fabrication	The body can be 3D-printed Magnets are available at low cost	Need of a custom manufacturing
Size	Can be designed with the size of a 6-wells plate or a 10-cm petri dish	Each coil is 40 × 86 mm (length × outer diameter)
Mounting on a microscope	Simple, given the flexible design options	Require a ≈ 140 × 95 mm rectangular hole in the microscope stage.
Generated field	With a simple array: 1 fixed field, from 1 to 90 mT. With a nested array: Tunable field, from 0 to 30 mT. The adjustments cannot be done live during an experiment	Field adjustable in live during the experiment by tuning the intensity of the current supplied to the coils. The field can go from 0 to 60 mT, but high magnitudes cannot be maintained too long, due to the Joule effect heating the coils

3.3 Super-paramagnetic Beads

1. The properties of the superparamagnetic beads are crucial to the feasibility and the precision of this technique. The size of the beads in particular is key: they need to be big enough to generate a sufficient dipolar force and for their position to be accurately determined with brightfield microscopy. We have

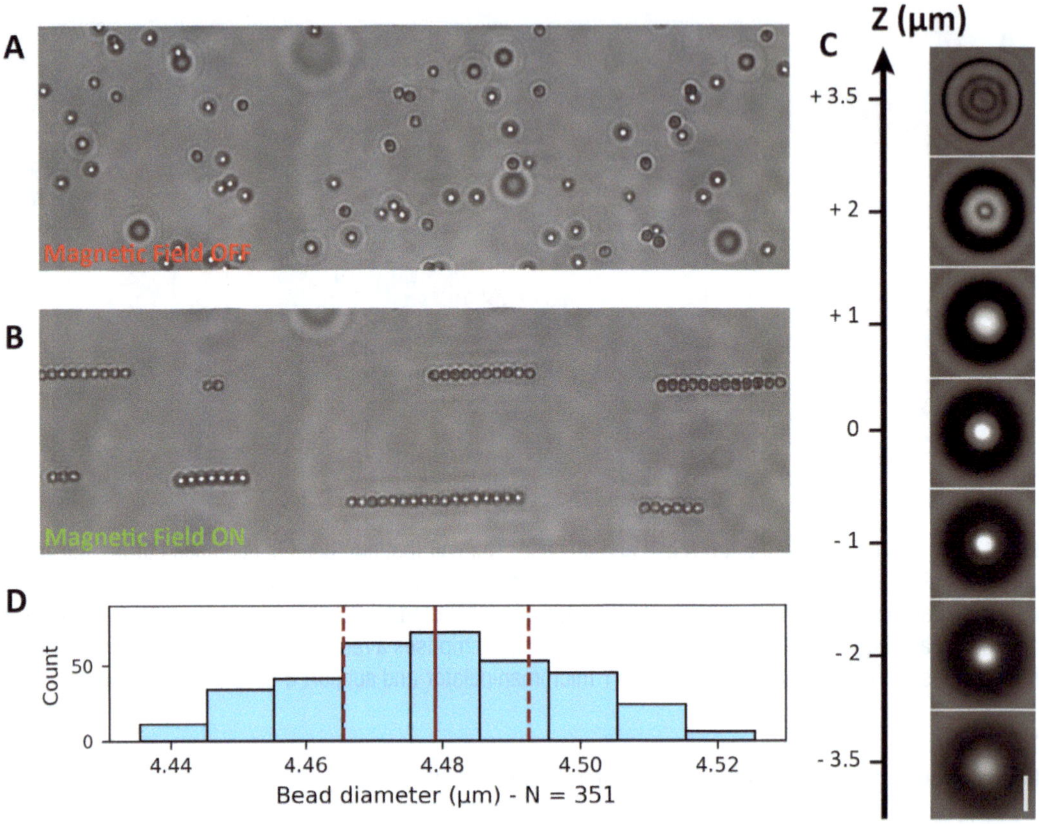

Fig. 5 M-450 Dynabeads. (**a**, **b**) M-450 Dynabeads before (**a**) and after (**b**) exposition to an external magnetic field. Without field, the beads undergo Brownian motion. When exposed to an external field, the beads magnetize and attract each other to form pairs or chain aligned with the field direction (horizontal here). (**c**) Aspect of a bead along the vertical direction (scale bar: 2 μm). The zero of the Z-axis is here the plane where the light spot observed below the bead is the brightest. In planes higher or lower, the intensity decreases progressively, and the shape of the light spot changes. Here, the pixel value range is the same on all the images but has been adjusted to increase the contrast. (**d**) Histogram of the bead size distribution within one batch of M-450 Dynabeads, with a bin size of 10 nm. The red line is the median,and the dashed lines are the quartiles. Over N = 351 beads, the average diameter is 4476 nm with a standard deviation of 19 nm

identified the M-450 Epoxy Dynabeads (Dynal, Thermo Fisher, USA) as the best choice since they ally many important features.

2. Beads from the Dynabead line are very monodisperse in diameter. A given batch of M-450 Dynabeads typically has an average diameter close to 4.5 μm with a standard deviation of less than 25 nm (Fig. 5d).
3. M-450 possess one of the highest density of magnetic particles [12] and the largest size of all Dynabeads. Thus, they can acquire high magnetic moment magnitudes, which result in a large range of pinching forces: from 1 pN to more than 1 nN. Moreover, like other Dynabeads, they have a very small residual

magnetization, which ensures that they stop attracting when the field is brought back to zero.

4. Optical properties: When observed under transmitted bright light illumination at high magnification, an intense light spot forms roughly 3.7 μm below the bead center (Fig. 5c). Computing the center of mass of this light spot, using the light intensity (pixel value) as a weight allows the center of the beads to be localized with a resolution that overcomes the diffraction limit: the precision of their localization in 2D is 2 nm and 10 nm in 3D [8, 13].
5. They are relatively easy to take up for cells. This depends on the cell type and can be tuned by coating the beads with diverse molecules, but every cell type considered for this experiment so far has proved able to take up M-450 beads.
6. This technique has also been performed successfully using M-270 beads, which are smaller (typical diameter: 2.7 μm) and magnetize less strongly. The method did not work when attempted with MyOne (typical diameter: 1 μm). The M-450 beads are nonetheless the best choice in terms of robustness and precision.
7. When using superparamagnetic Dynabeads, it is possible to determine the value of the attractive force one bead applies on its neighbor as a function of the external magnetic field and the bead positions. This way, the magnitude of the pinching force applied to the cortex can be computed, which is useful when interpreting the thickness measurements. However, since it is not essential for the use of this technique, we will not present this aspect here; the details are available in [8, 13].

3.4 Imaging System

To monitor the actin cortex thickness in time, a pair of beads pinching the cortex must be tracked in 3D. This is done by acquiring a time-lapse movie of the beads illuminated in bright field, with the focus on the light spot below the beads. To further improve the precision of the bead tracking, a Z-stack of three images is acquired at each time-point (see next section). A high magnification and a high dynamic range are required to produce images where the beads can be tracked precisely.

In total, in addition to the magnetic field generator, the imaging setup must include the following elements:

1. High-magnification objective (63× or 100×, preferably NA = 1.4).
2. Brightfield light source with a field diaphragm and a condenser diaphragm on the light path.
3. Camera with a 16-bits dynamic range. We use an Orca Flash4 (Hammamatsu, Japan).

4. Piezo element to control the position of the focal plane along the Z-axis. This can be a piezo controlled stage, or a PIFOC focus scanner. We use a PIFOC P-721.CDQ (Physik Instrumente, Germany).
5. Image acquisition software that can control both the camera and the Z-axis and save along with the images a log file containing the precise time of each image acquired. We use a custom-made Labview program, along with a data acquisition module (NI 6343, National Instruments, USA).
6. An environment control chamber to maintain the temperature of the sample to 37 °C throughout the experiment and ideally control the CO_2 concentration. We use The Box and The Cube (Life Imaging System, Switzerland).

3.5 Guidelines for Bead Imaging

Before imaging M-450 Dynabeads, the microscope and camera setting have to be adjusted following these principles:

1. The brightness of the light spot below the beads is maximal around 3.7 μm below the equatorial plane for M-450 Dynabeads. We call it the "plane of maximum intensity."
2. First focus on a bead in the plane of maximum intensity. Set the exposure time to 5 ms.
3. Set the Köhler illumination conditions on the microscope, then slightly open the field diaphragm to make the light spot sharper. Verify by moving the focal plane up and down. The light spot spreads a bit as you move away from the maximum intensity plane (Fig. 5c).
4. Adjust the light source power to use a significant fraction of the camera dynamic range. Typically, adjusting the maximum intensity to roughly 40,000 gray levels on a 16-bit camera is ideal. The image should never become saturated throughout the light spot.

When acquiring a time-lapse, the following points are keys to image the beads to allow a precise 3D tracking:

5. The tracking of the beads in the XY plane will be more precise when the focus is done on the plane where the light spot is the brightest (plane of maximum intensity).
6. The tracking of the beads along the Z-axis will be more precise when the focus is done on a plane located above or slightly below the plane of maximum intensity. This is because the shape of the light spot does not vary much when moving up and down from the plane of maximum intensity, but starts forming specific patterns when one move further below or—even better—above. These patterns facilitate the tracking along Z.

7. As mentioned before, the ideal solution to optimize tracking in both XY and Z is to acquire a small stack of three images for each time point. Typically, three planes 0.5 μm apart in Z and centered on the plane of maximum intensity will allow a very precise 3D localization of the beads.
8. If the conditions of imaging do not allow such Z-stack to be acquired at each time-point, the best plane to localize the bead precisely in XY and Z on a single image is one located slightly above the plane of maximum intensity, where the light spot below the bead starts spreading a little bit: it is the best compromise between points (1) and (2).

3.6 Experimental Execution

This protocol was optimized for two cell types: primary mouse dendritic cells (DC) and 3 T3 fibroblasts (3 T3). In the following protocol, some steps will differ according to which cell type they are applied to. This will be clearly stated. We believe that together these two cases are representative of a large number of cell types. Importantly, they encompass the cases of strongly adherent cells (3 T3) and weakly adherent ones (DC).

3.7 Bead Size Measurement

The distribution of diameters of the M-450 Dynabeads has a very low variance, with the exception of rare and obvious outliers, and varies only slightly from one batch to another. Nonetheless, for each vial of beads newly purchased, it is necessary to characterize precisely the average diameter, as it is one of the main sources of uncertainties of the technique. This is done easily using long chains of beads that form under specific conditions.

Image acquisition (*see* **Note 1**)

1. In a 35-mm glass-bottom petri dish, coat the glass surface with 1% bovine serum albumin (Merck, Germany) by incubating for 5 min. Rinse twice with PBS.
2. Adjust the volume of PBS contained in the dish to approximately 2 mL, and add 400,000 beads of the population you want to characterize. Rock gently the dish to homogenize as the beads sediment.
3. Place the dish in the experimental setup equipped with the magnetic field generator. Apply a field of 5–10 mT.
4. The beads will self-organize in long chains. Acquire several images of such chains in the plane of the beads light spot. Several criteria have to be fulfilled:

   As per usual, the light spot maximum intensity should be close to the maximum of the pixel range but never saturating.

   The chains must clearly be separated from each other and do not consider chains that are stacking on each other.

Within a chain, the bead aspect must be identical, indicating that they are all in the same plane along the Z-axis. If it is not the case, two explanations are probable. Either the dish is not horizontal enough, in which case it needs to be rectified. Or the beads are badly aligned, and in this case, one can switch off the field, mix the solution to resuspend the beads and switch the field on again during the beads sedimentation, if possible, with a high field value (30 mT).

5. Acquire images of chains to gather a total of several hundreds of beads.

Image Analysis

6. In the Fiji software, open the images of a chain of beads.
7. Use the method detailed below in Subheading 3.4.1. to detect the position of the center of each bead in the chain.
8. In the results table, use the columns "XM" and "YM"—the position in X and Y of the center of mass of each light spot—compute the series of neighbor-to-neighbor distance in the chain. Ordering the table row by XM value beforehand might simplify the task.
9. Repeat steps 6–8 for each of the acquired images.
10. Compute the mean and the standard deviation of the distribution of neighbor-to-neighbor distances. These values are used to characterize respectively the central tendency and the variability of the bead population.

3.8 Bead Preparation

Before using M-450 Dynabeads, it is necessary to rinse them from the stock solution medium and coat them with a molecule to tune the ability of cells to ingest them. Our standard coating protocol uses complete medium (with 10% FBS) to moderately increase the uptake rate by cells. Other options are fibronectin coating, to increase bead ingestion, and mPEG coating to decrease it (*see* **Note 2**).

Rinsing

1. The commercially available M-450 Dynabeads are conserved in distilled water with a concentration of 4×10^8 beads/mL. Vortex the stock solution vial to resuspend the beads and pipet 30 μL of the solution in an aliquot.
2. Add 1 mL of PBS to this aliquot and vortex the mix for 20 s.
3. Hold the aliquot vertically above a magnet (typically the W-10-N 1 cm neodymium cube from Supermagnete) for 20 s. The lower tip of the aliquot should be in contact with the magnet. The beads will sediment rapidly.

4. While holding the aliquot vertically above the magnet, gently remove the supernatant to disturb the bead pellet as little as possible.
5. Repeat steps 2–4 twice.

Coating with complete medium

6. In the aliquot containing your rinsed pellet of beads, add 100 μL of complete medium used for the culture of your cells of interest.
7. Vortex the mix for 30 s.
8. Before the beads sediment, place the aliquot on a rotating wheel for 3 h.
9. Conserve the aliquot at 2–8 °C, with the cap wrapped in Parafilm. If manipulated only in sterile conditions, it can last up to one month.

The resulting concentration expected in the aliquot is 120,000 beads/μL. Accounting for the loss of a fraction of the beads during the rinsing steps, the actual concentration should be around 100,000 beads/μL. To verify, one can dilute the solution 100-fold and count the concentration of beads with a typical cell-counting slide.

3.9 Experimental Chamber Preparation

While non-adherent or weakly adherent cells do not require a specific experimental chamber, adherent cells such as 3 T3 fibroblasts or macrophages do. Indeed, in the absence of surface treatment of their 2D substrate, those cell types would spread until they become flat, and their cortex undergoes large reorganization as the cells produce stress fibers. To prevent the cell to spread excessively, one needs to control the shape that the cells will adopt while adhering. This can be done by micropatterning a molecule that favors adhesion on the chamber's bottom side, such as fibronectin [14].

1. *Case of weakly adherent cells*

In a 35-mm glass-bottom petri dish, coat the glass surface with 1% bovine serum albumin (Sigma-Aldrich) by incubating for 5 min. Rinse twice with PBS.

Before adding the cells, rinse twice with warm imaging medium.

2. *Case of adherent cells*

This step is a variation on the technique described in [14]. Hence, we will not detail all the steps of the micropattering protocol, but rather how we adapted it to fit the constraints of our technique. Indeed, the original protocol used magnets to mount the coverslip on the microscope stage, which is obviously not possible here. Thus, we fabricate our own glass-bottom petri dish by assembling a holed dish and the micropatterned coverslip.

3. Petri dish preparation

We use 35-mm petri dishes with a 20-mm hole cut in the plastic bottom. Such dishes can either be bought or fabricated. To make them in the lab, we suggest to use a laser cutter, which can be programmed to cut a circular hole in a batch of precisely positioned petri dishes. One can custom make a guide to help with the positioning in the laser cutter.

4. Micropatterning on glass coverslip

We perform the aforementioned micropatterning protocol [15] on 25 mm round glass coverslips (thickness #1), using PLL-PEG to passivate the surface and a custom made photomask to burn the micropattern with deep UV light ($\lambda = 254$ nm). The micropatterned shapes are discs of identical diameters. For 3 T3 fibroblasts, we used 20-μm large discs. The diameter of the discs should be roughly equal to the typical diameter of the cell type used (when our 3 T3 cells are in suspension, they adopt a spherical shape with a typical diameter of 18 μm). Importantly, the micropatterned slides can be conserved several days, but the incubation with fibronectin should be done just before the experiment (*see* below).

5. Assemble the chambers

To stick the micropatterned coverslip on the bottom of the holed petri dish, use nontoxic silicon glue.

6. If they have been fabricated in the lab, the holed dish and its lid must be cleaned using 70% ethanol and dried thoroughly.
7. Pour some glue in a syringe (roughly 1 mL of glue for four dishes).
8. Place the holed dish upside down. Use the syringe to apply a thin and continuous ribbon of glue on the edge of the hole.
9. Using a plastic tweezer, place the coverslip on the glue circle, the micropatterned side facing down (toward the inside of the dish). Very gently press on the edges of the coverslip to push it against the dish and spread the glue. The circle of glue should be continuous so that the dish does not leak.
10. Let the assembled chamber dry overnight at 4 °C. At this stage, the assembled chamber can be conserved for approximately a week.
11. Coating with the protein of interest.

Because of the shorter conservation time of fibronectin, this step is only performed one day before or the day of the experiment (*see* **Note 3**).

12. Prepare a 10 μg/mL fibronectin solution in $NaHCO_3$ (pH 8.3) buffer.
13. Prepare a disc of Parafilm of 18 mm in diameter.
14. Pipet 150 μL of the fibronectin solution on the micropatterned glass surface of a dish.
15. Using a tweezer, place the Parafilm disc as a cap over the fibronectin droplet, in order to spread it on the entire surface of the coverslip as a uniform layer. Incubate for 20–30 min.
16. Without removing the Parafilm disc, add rapidly 2 mL of PBS in the dish. The disc of Parafilm should detach and float to the surface. Remove it and rinse twice more with warm imaging medium.

3.10 Cell Handling

Goal: Have the cell take up M-450 Dynabeads and transfer them to the experimental chamber. This part of the protocol is very dependent on the cell type used. We propose two options: one for the primary mouse dendritic cells (DC) that can take beads up fast and are not strongly adherent and another for the 3 T3 fibroblasts that take up beads less efficiently and are strongly adherent.

Dendritic Cells

Single step: Transfer to experimental chamber and incubation with beads (*see* **Note 3**).

1. In the BSA-coated petri dish, transfer 1 mL of cell in suspension in imaging medium at 2.5×10^5 cells/mL.
2. Mix 2.5 μL of the coated M-450 Dynabeads solution (*see* Subheading 3.8) with 0.5 mL imaging medium. Vortex and add to the chamber to obtain a 1:1 bead-to-cell ratio. Incubate at 37 °C, 5% CO_2 for 1 h.
3. After 1 h, roughly 25% of cells are expected to have ingested at least one bead. Verify that this is the case using a simple phase contrast microscope.
4. Optional: Only if the environment chamber on the microscope does not control CO_2. Add 0.5 mL of imaging medium supplemented with 20-mM HEPES.

3 T3 Fibroblasts

Incubation with beads (*see* **Note 4**)

1. 2 days before the experiment, seed approximately 2×10^5 cells in a 25-cm^2 culture flask.
2. Mix 6 μL of the coated M-450 Dynabeads solution (*see* Subheading 3.8) in 1 mL of warm culture medium. Vortex and add to the flask. Gently rock the flask to homogenize the distribution of beads.

3. After 2 days, a significant fraction of cells are expected to have ingested at least one bead (10–50% according to the bead preparation). Verify that hist is the case using a simple phase contrast microscope.

Transfer in experimental chamber (*see* **Note 3**)

4. On the day of the experiment, detach the 3 T3 cells from their flask with TrypLE and resuspend them in warm imaging medium.
5. In a micropatterned dish with 20-μm fibronectin discs (as prepared in Subheading 3.9) transfer 2 mL of cell suspension at 1.5×10^5 cells/mL. Let settle and adhere to the patterns for 20 min, at 37 °C 5% CO_2.
6. Check that most micropatterned discs are occupied by a cell that has started to adhere on the substrate.
7. *Flush* the bottom of the chamber to remove the non-adherent cells. To do this, hold a pipet in each hand and inject warm medium on one side as you aspire on the other side of the chamber. Never let the chamber dry. The flushing will generate a flux of medium in the chamber and efficiently remove excess cells, leaving only the adherent ones.
8. Let the cells adhere completely for 2 h at 37 °C and 5% CO_2.

Outer bead addition

9. Prepare a mix of 0.5 mL of warm imaging medium and 2.5 μL of the coated beads solution (*see* Subheading 3.8). Vortex and add to the dish.
10. Optional: Only if the microscope incubator does not control CO_2. Add 0.5 mL of imaging medium supplemented with 20-mM HEPES.

3.11 Magnetic Pincher Imaging

After placing the sample onto the microscope stage where the magnetic field generator is mounted, the first step is to acquire Z-stacks of beads. They will be used to compute a "Depthograph" (*see* Subheading 3.14 and Fig. 8). It is preferable to perform this step in the experimental dish containing the cells in the absence of a magnetic field: the formation of chains of beads would hinder the acquisition of clean Z-stacks. Then, the magnetic field can be applied to trigger the beads self-organization in the whole chamber, and many pinching events should occur simultaneously. For each imaged cell, acquire a 10 min time-lapse with a stack of images at different Z-positions for each time point (Fig. 6). Together with a log file indicating the precise time for the acquisition of each image, these movies are the raw data processed by the image analysis software to determine the cortex thickness as a function of time.

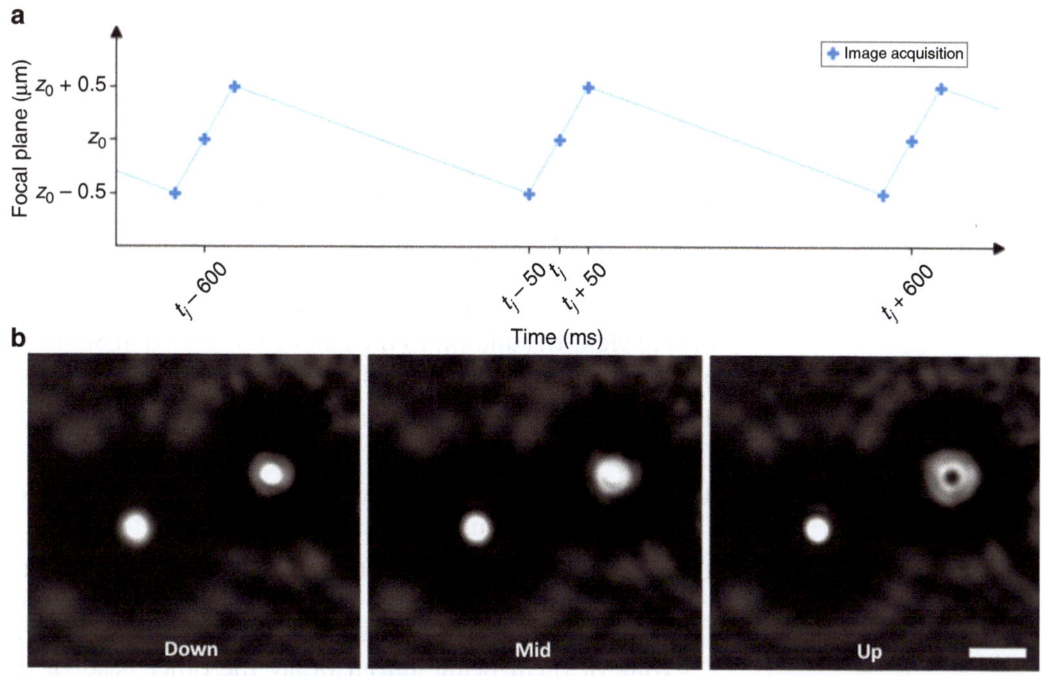

Fig. 6 Time-lapse acquisition. (**a**) Acquisition scheme. For each time-point ti, a stack of three frames is acquired. Those three frames are acquired every 50 ms and are distant of 0.5 μm along the Z-axis. (**b**) Example of such triplet of frames, taken from the same film as Fig. 1 (Scale bar: 2 μm). Note that here the two beads do not have the same Z coordinate: the left bead light spot is at its brightest in the "Up" frame, while for the right bead it is in the "Down" frame. The left bead is therefore slightly above the right bead. NB: The bright spot is saturated in this image for better visualization, but should not be saturated during the course of the experiment

Z-scan of beads (reference Depthograph)

1. Set the microscope light power and the field aperture as explained in Subheading 3.4. These setting should not change throughout the whole imaging session.
2. Position one or several beads in the field of view. These beads should be clearly separated from each other, perfectly still, and their image not affected by any object in the chamber or on the optical path. Position the focus in the equatorial plane of the beads (*see* Subheading 3.4).
3. Acquire a Z-stack of these beads: 401 steps every 20 nm, for a total course of 8 μm. The piezo element must be used to ensure precision of the displacement along Z.
4. Repeat steps 2 and 3 until at least eight Z-stacks of beads have been successfully acquired.

Magnetic Pincher on live cells (*see* **Notes 5** and **6**)

5. "Switch on" the magnetic field generator. This is done by either placing the Halbach array on its support, around the sample, or by sending current in the Helmholtz coils. The field magnitude should be between 3 and 10 mT, typically 5 mT, and be kept constant across experiments to obtain comparable results. It is strongly recommended to use a Gaussmeter to measure the applied field in situ.
6. Position a cell whose cortex is pinched by a pair of beads in the field of view. In addition to the guidelines defined in Subheading 3.5, the following criteria must be met:

   These beads should be as much as possible aligned along the magnetic field lines and in the same plane.

   The light spot below them should not be hindered by extracellular or intracellular objects.

   The pair of beads can be part of a longer chain that extend into, or out of the cell, or even both. However, beads pinching the cortex must have a maximum of two nearest-neighbors (one of them being automatically the other bead of the pair) that are well aligned with the line of magnetic field. Otherwise, the magnetic field in the region of the beads will not be properly defined.
7. Acquire a time-lapse of these beads (Fig. 6a, b):

   2–10 min, one z-triplet of images every 600 ms, for a total of 200 to 1000 time-points. Shorter durations allow for more movies of different cells to be acquired. Longer time-lapse movies allow for a more extended measurement of each cell cortex behavior.

   Each z-triplet is a short Z-stack of three images, with a step of 0.5 μm, ideally centered on the plane of maximum intensity. The piezo element must be used to ensure precision of the displacement along Z.

   Allow at least a delay of 50 ms between each step of the Z-stack, to ensure the precision of the positioning along Z.
8. Make sure that the software used for the acquisition also saves the precise time at which each image was taken. This can be attached to the images metadata or saved in a separate log file.
9. Repeat steps 2 and 3 to acquire a dataset representative of your cell population. A given chamber should not be images for longer than 2 h. Typically, 10–20 cells should be acquired in each chamber, depending on the duration of the acquired time-lapse movies.

3.12 Image Analysis: Overview

1. This section details the image analysis workflow used to extract the cortical thickness as a function of time from the acquired time-lapse images. It requires two different software packages: ImageJ (in our case Fiji [16]) and Python.
2. To analyze a group of time-lapse imaged from the same dish, the first task is to generate a Depthograph. It simply consists of a YZ-plane averaged profile of the bead, used to determine the relative positions of the beads along the Z-axis [17].
3. Then a simple tracking algorithm is applied to the time-lapse movies of Magnetic Pincher, to compute the trajectories of the beads in 3D. As many other tracking algorithms, ours proceed in two classic steps: (i) object segmentation and (ii) frame-to-frame matching. Step (i) is done semiautomatically with Fiji, while step (ii) has been automatized in a Python function.
4. The Python code can be found on Github: https://github.com/jvermeil-biophys/CortExplore_MIMB.git. The main libraries required to run the code are the following: os, time, numpy, pandas, scipy, scikit-image [18], matplotlib, and pyautogui.

3.13 Detect Bead Position in the XY Plane Using Fiji (See Note 7)

This detection method is routinely used across many of the image analysis workflows presented in this chapter. While simple, it allows a detection of the center of a bead in the XY plane with a very high precision. It consists in segmenting of the light spot below a bead, then computing its center of mass, and using the pixel value as weight (*see* Fig. 7).

1. In the Fiji software, open an image or a movie containing beads (Fig. 7a).
2. In "Set Measurement" select "Area," "Standard deviation," "Center of Mass," "Stack Position."
3. Use "Image/Adjust/Threshold" to manually segment the light spots under the beads (Fig. 7b, c). Enable the "Stack histogram" option. If the file analyzed is a movie, the same threshold value is used to segment the beads in all the frames of the movie. The regions segmented should cover each light spot, be roughly circular and have a diameter around 1–2 μm. Leave the "Threshold" dialog box open with the thresholded regions visible on the image.
4. Use the "Image Particle" tool to measure the properties of each region (Fig. 7d), with the following options enabled: "Display results," "Exclude on edges," "Include holes," "Clear results." Adjust the area and circularity criterion so that the only objects analyzed are the segmented light spots. Select "Show: Outlines" to visually check that all the light spots were analyzed.
5. Save the Results table (Fig. 7e) in .txt format.

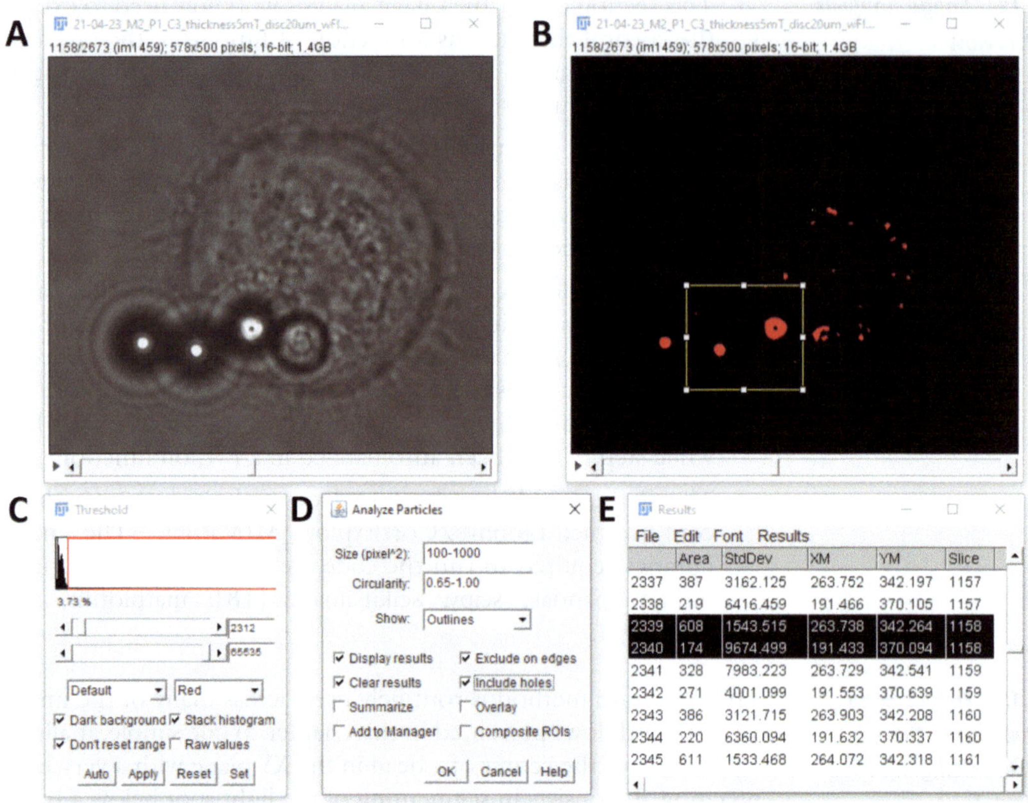

Fig. 7 Localize bead centers with Fiji. (**a**) Raw frame of a Magnetic Pincher time-lapse. Only the contrast has been modified. The image has no scale (results in pixel values). (**b**) The same frame, where the threshold has been applied and a rectangular region of interest has been drawn around the beads pinching the cortex. (**c**) The "Threshold" dialog box. It has been manually adjusted to segment properly the beads light spot in all frames of the time-lapse. (**d**) The "Analyze Particles" dialog box. The parameters shown here are the set typically applied. (**e**) The "Results" windows that open after running "Analyze Particles." The two selected lines correspond to the two light spots segmented inside the ROI in (**b**). The columns XM and YM contain the coordinates of the centers of mass of the light spots (center weighted by the pixel values). Note that, here, running "Analyze Particles" with the option "Show: Outlines" would also open an "Outlines" window (not shown here) useful to assess the quality of the detection

3.14 Make a Depthograph from the Bead Z-Stacks (See Note 8)

In order to make a reference Depthograph, process bead Z-stacks acquired for a given experiment to compute a typical YZ profile of the beads.

Detection of the beads center, in Fiji.

1. In the Fiji software, open a Z-stack of one or several beads (acquired as specified in Subheading 3.11). For each bead crop, a small rectangular region of interest around the bead and save the resulting Z-stack in .tif format.
2. Open one of these single-bead Z-stack.

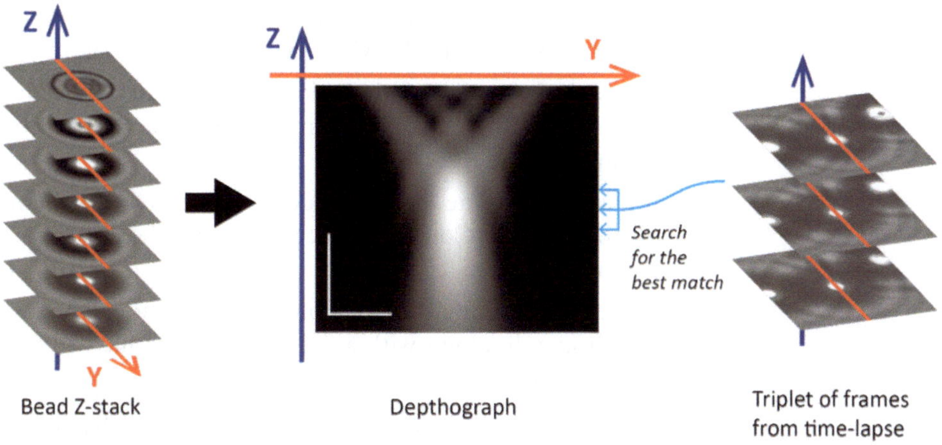

Fig. 8 Depthograph. Computation and application of the Depthograph. On the left, a Z-stack used to generate the Dephtograph. In the center, an example of such Depthograph, with the depth (Z) on the vertical axis and the profiles (orange lines) on the horizontal axis (vertical scale bar: 2 μm; horizontal scale bar: 1 μm). On the right, a typical application of the Depthograph: to locate the bead along the Z-axis, one can take its profiles (orange lines) on the frames of a Z-triplet and compare them with every row of the Depthograph to find the best match. This approach, using the fixed distance between the three frames (0.5 μm) as an additional information, ensures the uniqueness of the best match and improve the precision. Therefore, each bead can be located within a common reference Depthograph, and the distance ΔZ between the beads pinching the cortex can be computed

3. Use the method detailed in Subheading 3.13. to detect the position of the center of the bead on most frames of the Z-stack. Not managing to detect the light spot on the first and the last frames is normal. The light spot should be successfully segmented typically from frame 100 to frame 300 over a total of 401 in the stack, without any misdetection within this range. Check visually with the "Outlines" that it is the case.
4. Save the Results table in .txt format in the same folder as the corresponding .tif single-bead Z-stack. By convention, the name of this .txt file should be the same as the .tif file, appended with "_Results.txt."

Computation of the Depthograph, in Python.

This task is performed using the Main_DepthoMaker.py script from the CortExplore_MIMB package. In order to run it follow these steps:

5. Gather all the .tif bead Z-stack files for a given experiment in a folder. Each .tif file must be accompanied with the corresponding "_Results.txt" file generated in the previous step. The naming convention must be strictly respected.
6. Open the Main_DepthoMaker.py script in any development environment (e.g., Spyder). Run the line importing the depthoMaker() function from SimpleBeadTracker.py. If it

returns an error, check that the working environment is defined as the "CortExplore_MIMB" folder (where the .py files are).

7. Indicate the paths of the relevant directories by filling the dictionary "dictPaths" according to the instructions in comment.
8. Indicate the value of the relevant parameters by filling the dictionary "dictConstants" according to the instructions in comment.
9. Run the line calling the depthoMaker() function. The resulting Depthograph, along with a "_Metadata.csv" file will be saved in the specified directory.

The precise instructions to run this code are detailed in the Github repository documentation.

As a reference, here are the outlines of the functioning of the depthoMaker() function.

Open a .tif Z-Stack of a bead. Using the previously computed Results file, on each frame, translate the image so that the center of mass of the bead is exactly in the center of the image. This is done using skimage transform.warp() function with a bilinear interpolation.

On each frame, take the vertical intensity profile of the bead. This can be done by averaging the five central vertical lines of the frame. This intensity profile is a 1D array of pixel value.

Concatenate all these arrays along a second dimension to obtain a 2D array of intensity profiles. The first dimension is Y, and the second is Z. This constitutes a Depthograph for a given bead.

Repeat the steps 1–3 for each stack of beads corresponding to an experiment. Average all the Depthographs, by using the plane of maximum intensity to align them. The resulting 2D array is the average YZ profile of beads imaged to produce the Depthograph of a given experiment.

3.15 Segment the Beads with ImageJ

As a first step to track the beads positions in the Magnetic Pincher time-lapse, segment the light spots of the beads of interest in all the frames in ImageJ/Fiji and determine the position of their center.

1. In the Fiji software, open a Magnetic Pincher time-lapse. If necessary, convert it to .tif format.
2. If the images are organized as Z-T hyperstacks, flatten to obtain a one-dimensional stack where images are ordered in the following way: (t1, z1), (t1, z2), (t1, z3), (t2, z1), (t2, z2), (t2, z3), . . ., (tN, z1), (tN, z2), (tN, z3), where t1 . . . tN are all the successive time-points and z1, z2, z3 are the lower, middle, and upper Z-planes, in that order. To do so, use the function "Image/Hyperstack/Hyperstack to Stack."

3. Use the method detailed in Subheading 3.13. to detect the position of the center of each bead of interest in the time-lapse. Check visually with the "Outlines" plot that the light spots corresponding to the beads of interest were analyzed in nearly all the frames. Save the Results table in .txt format. By convention, the name of this .txt file should be the same as the .tif file, appended with "_Results.txt."
4. Repeat steps 1–3 for each of the acquired Magnetic Pincher time-lapse.

3.16 Track the Beads in 3D with a Custom-Made Python Algorithm (See Notes 9 and 10)

The second step of the tracking is performed using the Main_3-DTracker.py script from the CortExplore_MIMB package.

1. Gather all the .tif Magnetic Pincher time-lapses files for a given experiment in a folder. Each .tif file must be accompanied with two other files: (i) the corresponding "_Results.txt" file generated in the previous step and (ii) a file containing the precise date in millisecond of each image as a single text column. By convention, the name of this .txt file should be the same as the .tif file, appended with "_Timepoints.txt."
2. Open the Main_3DTracker.py script in any development environment (e.g., Spyder). Run the line importing the mainTracker() function from SimpleBeadTracker.py. If it returns an error, check that the working environment is defined as the "CortExplore_MIMB" folder (where the .py files are).
3. Indicate the paths of the relevant directories by filling the dictionary "dictPaths" according to the instructions in comment.
4. Indicate the value of the relevant parameters by filling the dictionary "dictConstants" according to the instructions in comment.
5. If necessary, set optional parameter values by filling the dictionary "dictConstants" according to the instructions in comment.
6. Run the line calling the depthoMaker() function.
7. For each Magnetic Pincher time-lapses file, the program will start by displaying the first frame, where detected beads labeled with orange crosses. If both of the beads pinching the cortex are properly detected on this first image, click on "Yes" and then click on the positions of the two beads (the clicks on the image do not have to be very precise). If one of the two beads is not detected, click on "Next Frame" and continue.
8. As the program matches the positions of the beads frame by frame to build their trajectories, it might happen that the algorithm lose track of one of the beads of interest, either because it was not detected on a given frame, or because its

frame-to-frame motion was significant. Then, the program displays this frame. Once again, if both of the beads pinching the cortex are properly detected on this first image, click on "Yes" and then click on the positions of the two beads (the clicks on the image do not have to be very precise). If one of the two beads is not detected, click on "No."

9. Once the program have built the trajectories, it will ask the user to assess for each of the beads of interest whether it is in or out of the cell (useful to save the trajectories) and whether it has one or two neighbors (useful to compute the pinching force).
10. [Automatic step] After these steps, building the trajectories of the beads in the XY plane, the program automatically computes the motion of the beads along the Z-axis by comparing for each frame the vertical profile of the beads of interest to the Depthograph generated for this experiment. This step can be time-consuming.
11. [Automatic step] From the XY and Z positions of the beads, the program computes the 3D center-to-center distance between the beads.
12. [Automatic step] The program computes the force applied by the beads on the cortex for each time-point, using the method detailed in [8].
13. [Automatic step] Finally, the program saves a "_timeseries.csv" file in the specified folder.

The resulting "_timeseries.csv" file is a table containing the evolution with time (column "T") of different quantities: the beads center-to-center distances along each axis ("dx," "dy," "dz"), in the XY-plane ("D2") and in three dimensions ("D3"). It also contains the magnetic field ("B," constant) and the force ("F," varying due to the fluctuations in distance). To compute the time-resolved thickness of the cortex, subtract the beads average diameter (*see* Subheading 3.7) to the distance in three dimensions ("D3"). Average thickness of the cortex and fluctuation amplitude can be readily obtained from this time-series (*see* **Note 11**).

3.17 Discussion

1. The Magnetic Pincher can probe the actin cortex thickness in live cells and monitor its fluctuations with high precision and time resolution. One of the key features of this technique is its flexibility: the version using the Halbach array to generate the field can easily be installed on any microscope, and the requirements regarding the imaging setup are easy to fulfil (simple bright field illumination). It is also surprisingly easy to apply to many different cell types: while one might think that having nonimmune cells to ingest 4.5-μm large beads is a hard task, some adjustments to the beads coating (*see* Subheading 2) can make it very robust. As of today, this technique has been

successfully applied not only to mouse primary dendritic cells and 3 T3 fibroblasts but also to *Dictyostelium amoeba*, to HoxB8 immortalized macrophages, to RPE1 human retinal pigment epithelial cells, and to MDCK epithelial cells. It is also simple to use the Magnetic Pincher in combination with other classic tools in cell biology: chemical drugs, epifluorescence imaging, or optogenetic systems.

2. An important point to be made regards the definition of the cortical thickness we measure with the Magnetic Pincher. As recent observations have shown, the actin cortex is not a homogeneous layer: the density of actin [19] and other components (e.g., myosin II [20]) varies along the cortex depth. It is also connected to the rest of the actin cytoskeleton. Here, we define the thickness of the actin cortex based on its mechanical properties: it is defined as the layer that withstand the small stress applied by the beads pinching it. This means the measured value is not exactly the thickness as it could be defined based on actin density. Future developments should allow to compare the various measures. The Magnetic Pincher-based definition is convenient when considering the mechanical function of the actin cortex, because a fundamental attribute of the cortex is its density in branched and cross-linked actin filaments, which can withstand the applied stress, as opposed to the more loosely connected filaments present in the cytoplasm.
3. A natural extension of this technique consists in using the beads to indent the cortex more significantly by applying a controlled force. This way, by tracking the beads to compute the cortex deformation in response, we can compute mechanical characteristics of the cortex, such as an elastic modulus [8]. The cortex being a very complex material (viscoelastic, nonlinear, active, and nonuniform) interpreting this type of measurement is challenging, and their in-depth characterization is still under development. Yet the Magnetic Pincher, which can apply a large range of forces directly on the cortical layer, seems like a promising tool for these rheology experiments.

4 Notes

1. Bead imaging

If the bead appearance in trans-illumination seems very different than on Fig. 5C, one should adjust the field diaphragm and light source intensity. Setting the Köhler conditions is a good starting point. The "dotty" effect due to the pinholes when using a spinning disc microscope in brightfield mode can be problematic (this might be the case when performing simultaneous fluorescence imaging using such a microscope).

2. Bead uptake by the cells

The uptake of M-450 Dynabeads by cells is a crucial step that affects a lot the overall success and throughput of the technique. It is recommended to optimize carefully this step when using the technique for the first time or when applying it to a new cell type. We suggest two alternative coating options (*see* Subheading 1.3.4) to increase or decrease the affinity of cells for beads. Increasing it is useful to ensure enough beads are ingested. Decreasing it is useful for outer beads if the cells are very prone to ingest them during time-lapse acquisitions.

- To increase bead uptake: coat the beads with fibronectin.
- To decrease bead uptake: coat the beads with mPEG.

With fibronectin

The protocol is the same than with complete medium (*see* Subheading 1.3.4) but replacing the medium in step 1 by 100 μL of fibronectin diluted in NaHCO3 (pH 8.6) at 25 μg/mL (*see* Materials section).

With mPEG

Beads with a less-adhesive surface can be made by using the Dynabeads included in the CELLection™ Biotin Binder Kit (11533, Thermo Fisher, USA). These are M-450 Dynabeads functionalized with streptavidin via a DNA linker, which allows to bind biotinylated proteins to the surface.

Follow the rinsing protocol mentioned above (*see* Subheading 1.3.4) but replace the complete medium in Step 2 with a solution of mPEG(5 K)-Biotin (JKA3097, Merck, Germany) in HEPES (10 mM, pH 7.4) at a concentration of 1 mg/mL.

3. Micropatterned chamber manufacturing

For adherent cells that require a micropatterned substrate, the experimental chamber manufacturing can prove challenging. To adopt a "quality-control" approach when optimizing this step, it is useful to make the micropatterns fluorescent, by adding 3 μg/mL of labeled fibrinogen in the fibronectin solution (Fibrinogen from Human Plasma, Alexa Fluor Conjugate, Thermo Fischer, USA).

4. Assess if a bead is inside a cell

To ensure a bead is inside a cell at high magnification, focus on the equatorial plane of the bead (Fig. 2e). At low magnification, gently moving the experimental chamber to induce sloshing can help to distinguish the floating beads from the engulfed ones.

5. Difficulty to find an adequate pinching event

In such case, one can adjust the cell-to-beads ratio suggested in Subheading 3.10. If many cells have a bead inside but none in their close environment, one can add additional beads during the imaging, or induce some gentle motion of the experimental chamber to make free beads move.

6. Motion of the cells during imaging

Depending on cell types, the cell activity can be very likely to cause a motion of the beads in X, Y and Z. In this case, one should follow the beads with the stage and with the focus to ensure the imaging conditions of the beads stay as good as possible throughout the acquisition: as the thickness is computed from the beads relative distance, their absolute position in the field of view is not important.

7. Difficulty when segmenting beads

In the method described in Subheading 3.13, the thresholding should ideally be stringent and allow to segment only the beads light spots. However, it is often not possible, for instance, because the beads moved up or down during the acquisition. Then a solution is to increase significantly the threshold value and define a rectangular ROI around the beads, before running the "Analyze Particle" tool.

8. Difficulty when making a Depthograph

If Depthographs look noisy or discontinuous, it might be because of one of the Z-stacks used to compute it. Select only the good beads for those Z-stacks: as fixed to the substrate as possible. Do not hesitate to rule them out a posteriori if a given Z-stack gives a very noisy Depthograph.

9. Difficulty when tracking the beads

If one of the beads of interest is not detected on too many frames of a time-lapse, the tracking step in python will take a very long time. Do not hesitate to re-do the segmentation step described in Subheading 3.13 to try to improve the bead detection. Or adjust the imaging parameters for next experiment.

10. Tracking along the Z-axis

One of the main difficulties with the deployment of the Magnetic Pincher comes from the tracking of beads along the Z-axis: the fact that it requires stacks of three images at different Z adds complexity to the setup. This may be solved in future developments of the technique, either using deep learning to better analyze the features of the images, or improving the setup with quantitative phase imaging to obtain an additional source of information to track the beads along the Z-axis [21]. Both of these upgrades could enable the tracking of microbeads based on a single Z-plane.

11. Throughput

Since this technique involves a magnification that does not usually allow several pinched cells to be observed in the same field of view, it does not have a very high throughput: within a day of experiment, a few tens of cells can be probed. This can be mitigated by acquiring shorter time-series.

References

1. Chugh P, Paluch EK (2018) The actin cortex at a glance. J Cell Sci 131:jcs186254. https://doi.org/10.1242/jcs.186254
2. Fritzsche M, Lewalle A, Duke T et al (2013) Analysis of turnover dynamics of the submembranous actin cortex. Mol Biol Cell 24:757–767. https://doi.org/10.1091/mbc.e12-06-0485
3. Svitkina TM, Neyfakh AA, Bershadsky AD (1986) Actin cytoskeleton of spread fibroblasts appears to assemble at the cell edges. J Cell Sci 82:235–248. https://doi.org/10.1242/jcs.82.1.235
4. Chugh P, Clark AG, Smith MB et al (2017) Actin cortex architecture regulates cell surface tension. Nat Cell Biol 19:689–697. https://doi.org/10.1038/ncb3525
5. Charras GT, Hu C-K, Coughlin M et al (2006) Reassembly of contractile actin cortex in cell blebs. J Cell Biol 175:477–490. https://doi.org/10.1083/jcb.200602085
6. Eghiaian F, Rigato A, Scheuring S (2015) Structural, mechanical, and dynamical variability of the actin cortex in living cells. Biophys J 108:1330–1340. https://doi.org/10.1016/j.bpj.2015.01.016
7. Clark AG, Dierkes K, Paluch EK (2013) Monitoring actin cortex thickness in live cells. Biophys J 105:570–580. https://doi.org/10.1016/j.bpj.2013.05.057
8. Laplaud V, Levernier N, Pineau J et al (2021) Pinching the cortex of live cells reveals thickness instabilities caused by myosin II motors. Sci Adv 7:eabe3640. https://doi.org/10.1126/sciadv.abe3640
9. Pierre P, Turley SJ, Gatti E et al (1997) Developmental regulation of MHC class II transport in mouse dendritic cells. Nature 388:787–792. https://doi.org/10.1038/42039
10. Halbach K (1985) Application of permanent magnets in accelerators and electron storage rings (invited). J Appl Phys 57:3605–3608. https://doi.org/10.1063/1.335021
11. Tretiak O, Blümler P, Bougas L (2019) Variable single-axis magnetic-field generator using permanent magnets. AIP Adv 9:115312. https://doi.org/10.1063/1.5130896
12. Fonnum G, Johansson C, Molteberg A et al (2005) Characterisation of Dynabeads® by magnetization measurements and Mössbauer spectroscopy. J Magn Magn Mater 293:41–47. https://doi.org/10.1016/j.jmmm.2005.01.041
13. Bauër P, Tavacoli J, Pujol T et al (2017) A new method to measure mechanics and dynamic assembly of branched actin networks. Sci Rep 7:15688. https://doi.org/10.1038/s41598-017-15638-5
14. Azioune A, Storch M, Bornens M et al (2009) Simple and rapid process for single cell micro-patterning. Lab Chip 9:1640. https://doi.org/10.1039/b821581m
15. Carpi N, Carpi N, Piel M et al (2011) Micropatterning on glass with deep UV. Protoc Exch. https://doi.org/10.1038/protex.2011.238
16. Schindelin J, Arganda-Carreras I, Frise E et al (2012) Fiji: an open-source platform for biological-image analysis. Nat Methods 9:676–682. https://doi.org/10.1038/nmeth.2019
17. Gosse C, Croquette V (2002) Magnetic tweezers: micromanipulation and force measurement at the molecular level. Biophys J 82:3314–3329. https://doi.org/10.1016/S0006-3495(02)75672-5
18. Van Der Walt S, Schönberger JL, Nunez-Iglesias J et al (2014) Scikit-image: image

processing in python. PeerJ 2:e453. https://doi.org/10.7717/peerj.453

19. Clausen MP, Colin-York H, Schneider F et al (2017) Dissecting the actin cortex density and membrane-cortex distance in living cells by super-resolution microscopy. J Phys Appl Phys 50:064002. https://doi.org/10.1088/1361-6463/aa52a1
20. Truong Quang BA, Peters R, Cassani DAD et al (2021) Extent of myosin penetration within the actin cortex regulates cell surface mechanics. Nat Commun 12:6511. https://doi.org/10.1038/s41467-021-26611-2
21. Bon P, Bourg N, Lécart S et al (2015) Three-dimensional nanometre localization of nanoparticles to enhance super-resolution microscopy. Nat Commun 6:7764. https://doi.org/10.1038/ncomms8764

Chapter 11

Measurement of Forces Acting on Single T-Cell Receptors

Lukas Schrangl, Janett Göhring, Florian Kellner, Johannes B. Huppa, and Gerhard J. Schütz

Abstract

Molecular forces are increasingly recognized as an important parameter to understand cellular signaling processes. In the recent years, evidence accumulated that also T-cells exert tensile forces via their T-cell receptor during the antigen recognition process. To measure such intercellular pulling forces, one can make use of the elastic properties of spider silk peptides, which act similar to Hookean springs: increased strain corresponds to increased stress applied to the peptide. Combined with Förster resonance energy transfer (FRET) to read out the strain, such peptides represent powerful and versatile nanoscopic force sensing tools. In this paper, we provide a detailed protocol how to synthesize a molecular force sensor for application in T-cell antigen recognition and hands-on guidelines on experiments and analysis of obtained single molecule FRET data.

Key words Single molecule microscopy, Förster resonance energy transfer (FRET), T-cell, Immunological synapse, Molecular forces, Force sensor, Single-molecule tracking

1 Introduction

In the recent years, it has become increasingly clear that mechanical forces are pivotal for cellular signaling processes [1]. Apart from serving as readout for mechanosensors—that is, proteins that change conformation when put under load [2]—forces may also affect the kinetics of ligand-receptor interactions [3–7]. For example, in cellular contact regions, distance fluctuations between the two opposing cell surfaces naturally put the intercellular junctions under tensile stress, thereby affecting bond durations. We and others have recently found that forces are of particular interest in the process of T-cell antigen recognition [3, 4, 8–11]. Antigen recognition proceeds in the highly organized interface between a T-cell and an antigen-presenting cell, termed immunological synapse [12]. During the recognition process, T-cells not only pull on activating surfaces [9, 11, 13–16]; also—in reverse—exertion of

Christoph Wülfing and Robert F. Murphy (eds.), *Imaging Cell Signaling*, Methods in Molecular Biology, vol. 2800,
https://doi.org/10.1007/978-1-0716-3834-7_11,

forces on the T-cell antigen receptor (TCR) triggers downstream signaling events [8, 10].

Quantification of cellular forces is commonly performed with one of the two following approaches: first, in traction force microscopy, the impact of cells moving on nanopillars or on hydrogels containing tracer beads is determined via measurements and quantitative analysis of the nanopillar bending or bead displacements, respectively [17, 18]. The two techniques were applied to measure global T-cell-exerted forces during the antigen-recognition process [13, 19, 20]. Since in this approach the antigen-presenting cell needs to be replaced by the functionalized surface, it comes with the disadvantage of providing rather limited flexibility in emulating the surface of the opposing cell. Moreover, it results in average force maps, without quantitative insights into the single-molecule pulling events. The second group of approaches utilizes Förster resonance energy transfer (FRET) or quenching assays to quantify the nanometer stretching of molecular springs such as the spider silk peptide [21, 22] or DNA hairpin structures [23]. Such molecular force sensors allow greater flexibility in their application, as they can be mounted directly in the immunological synapse [15, 24]. In addition, information about forces acting on single TCR molecules can be obtained.

Recently, we have reported the development and application of a single-molecule FRET-based force sensor to quantify tensile forces during T-cell antigen recognition [15]. Figure 1a shows the overall idea of our approach. The actual force-sensing unit is based on a spider silk peptide, to which we site specifically conjugated a FRET donor and acceptor (*see* Subheading 3.1 for more details). Force-imposed stretching of the spring unit hence leads to a decrease in FRET efficiency, which can be quantified, for example, via sensitized emission measurements. Since the spring unit has previously been calibrated [25], the obtained FRET efficiency can be converted into force values (*see* Fig. 1b). The sensor is anchored to a glass-supported lipid bilayer, which serves as a surrogate APC for antigen-specific or TCR-mediated T-cell stimulation [12]. The spring unit is tethered to either a peptide-loaded MHC acting as natural TCR ligand or a TCR-reactive single-chain antibody fragment, which provides a stable TCR-interaction. In such a fashion, this force sensor allows for quantifying the pulling exerted by a single TCR at a force resolution of ~1 pN and a time resolution of a few milliseconds. Below, we describe the synthesis of the sensor, and we provide protocols how to perform the according experiments and data analysis.

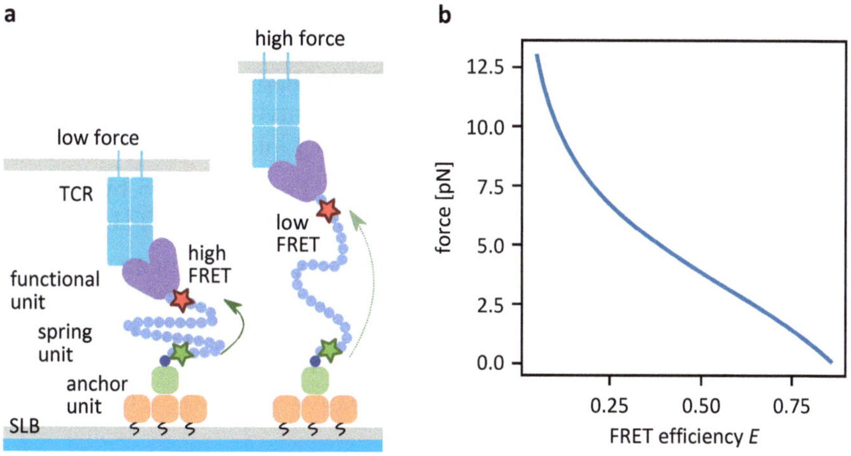

Fig. 1 The force sensor. (**a**) To bind the force sensor to a planar glass-supported lipid bilayer (SLB), we used a 3×His_6-tagged monovalent streptavidin (mSav-3×His_6) as anchor unit. The sensor's functional unit shall recognize the TCR, so that a T-cell contacting the SLB is able to tension the spring unit located between anchor and functional unit. We used a spider silk peptide as spring unit, which we labeled with a FRET donor and acceptor close to the anchor and functional unit, respectively. If the cell exerts low force, the spring unit is collapsed, resulting in high FRET efficiency between the fluorophores. Increased forces stretch the spring element and lead to reduced FRET efficiency. (**b**) The calibration curve allows for converting measured FRET efficiencies to forces. The functional relationship between the end-to-end distance of the spring and the applied force has been previously quantified [15, 25]. The depicted force-versus-FRET efficiency curve is the result of combining this relation with the well-known dependence of the FRET efficiency on the donor–acceptor distance.

2 Materials

2.1 Reagents

1. Imaging buffer: Hank's Balanced Salt Solution with calcium and magnesium +2% Fetal Calf Serum (FCS). *See* **Note 1** about FCS.
2. Fura-2: 1 mg/mL of Fura-2, AM dye (Thermo Fisher Scientific) in dimethyl sulfoxide.

2.2 FRET-Based Single-Molecule Force Sensors

1. Devices: FPLC with Superdex 200, Superdex 75 columns (GE Healthcare), and a MonoQ 5/50 column (GE Healthcare), HPLC (5-μm particle size, 250 × 21.2 mm, Agilent).

2.3 T-Cell Culture

1. T-cell medium: RPMI 1640 supplemented with 10% FCS, 100 U/mL penicillin/streptomycin, 2-mM glutamate, 1-mM sodium pyruvate, 1× nonessential amino acids, and 50-μM ß-Mercaptoethanol.

2.4 Supported Lipid Bilayers

1. Lipids: DPPC (1,2-dipalmitoyl-sn-glycero-3-phosphocholine), POPC (1-palmitoyl-2-oleoyl-glycero-3-phosphocholine), and Ni-NTA-DGS (1,2-dioleoyl-sn-glycero-3-[(N-(5-amino-1-carboxypentyl)iminodiacetic acid)succinyl] (nickel salt)) (Avanti Polar Lipids, Inc., USA)
2. Vesicles: Glass test tubes (Fiolax 12 × 75 by Schott AG, Germany), ultrasound water bath (USC500TH, VWR)
3. Slides used for microscopy experiments: Glass coverslips (MENZEL-Gläser Deckgläser 24 × 60 mm # 1.5), 8-well chambers (Nunc Lab-Tek Chambered Coverglass, Thermo Fisher Scientific), duplicating silicone (twin-sil extrahart, picodent, Germany), plasma cleaner (PDC-002, Harrick Plasma, USA)
4. His-tagged proteins: Mouse ICAM-1 and B7-1 with C-terminal His tags (Sino Biological, China)

2.5 Microscopy

See Fig. 2a for a schematic of the beam paths.

1. Light sources: 640-nm laser (OBIS 640, Coherent, USA), 532-nm laser (LCX-532L with L1C-AOM, Oxxius, France), 405-nm laser (iBeam Smart 405-S, Toptica Photonics AG, Germany). All lasers are pigtailed and fed into a beam combiner (OBIS Galaxy, Coherent). The beam from the output fiber is collimated (RC08APC, Thorlabs), then expanded, and focused into the back focal plane of the objective using two lenses (AC254-100-A-ML and AC508-300-A-ML, Thorlabs) in a telescope configuration. Shifting the beam parallel to the optical axis toward the edge of the objective enables total internal reflection fluorescence (TIRF) microscopy.
2. Microscope: Inverted microscope (Axio Observer 7, Zeiss) with a high-NA objective (α Plan-FLUAR 100x/1.45 Oil or α Plan-Apochromat 63x/1.46 Oil Corr, Zeiss) and a quad-band dichroic mirror (Di01-R405/488/532/635-25×36, Semrock Inc., USA) to reflect excitation light toward the sample while transmitting emission light toward the camera. Optics for epi-illumination have been removed from the microscope to enable TIRF illumination as described above. Using a 1× tube lens with 100× objectives and a 1.6× tube lens with 63× objectives results in an effective 100× magnification.
3. Emission: A beam splitter system (Optosplit II, Cairn Research, UK) allows for side-by-side projection of FRET donor and acceptor emission onto the EM-CCD chip of a camera (Andor iXon Ultra 897, Andor Technology Ltd., UK). To this end, a dichroic mirror (640 nm cutoff wavelength; FF640-FDi01-25×36, Semrock Inc., USA) is inserted in conjunction with bandpass filters for the respective channels (ET570/60m and ET675/50m, Chroma Technology Corp,

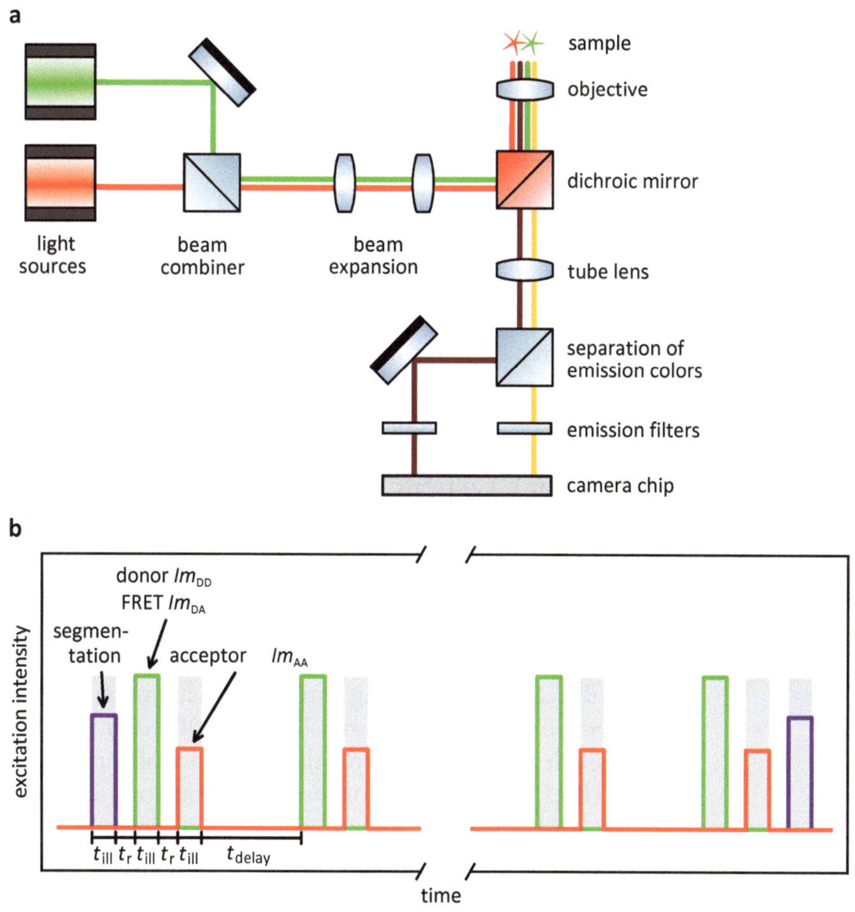

Fig. 2 Microscopy. (**a**) Beam paths for multicolor excitation of fluorophores and simultaneous imaging of their emission. See Subheading 2.5 for further information. (**b**) Typical illumination sequence. 405-nm excitation is indicated by the purple line, 532-nm excitation by the green line, 640-nm excitation by the red line, and the camera trigger by the gray shade. Labels and arrows indicate the types of images recorded during each illumination. See Subheading 3.5 for details

USA). Longitudinal chromatic aberrations can be corrected by placing a lens into one of the emission beam paths.

4. Control: Synchronized stroboscopic fluorophore excitation and camera readout are achieved via function generator hardware (PCI-6713, National Instruments, USA) in conjunction with our home-brew *SDT-control* software.

5. Fiducial markers for image registration: TetraSpeck beads, 200-nm diameter (well number 4 of TetraSpeck Fluorescent Microspheres Size Kit, Invitrogen).

2.6 Software

1. Python distribution: *Anaconda* with *conda-forge* channel enabled. To this end, installation of *miniforge* [26] is recommended.

2. Required Python packages: opencv, trackpy, ipympl, scikit-learn, pyqt, sdt-python, jupyterlab.
3. *git* version control system.

3 Methods

3.1 Synthesis of Force Sensors

1. **Anchor unit: Monovalent streptavidin with 3His$_6$ tags (mSav-3xHis$_6$)** Expression of monovalent streptavidin in BL-21 *E. coli* inclusion bodies using the pET expression system: Prepare one culture with the *dead* subunit (biotin binding deficient subunit equipped with a C-terminal His$_6$) and one culture with the functional *alive* subunit (biotin binding subunit with a C-terminal 3C protease cleaving site followed by Glu$_6$-tag). Cultivate cells at 37 °C while shaking. Induce culture using 1-mM IPTG at OD 0.4–0.6. Harvest cells approximately 4 h later or if expression profile is satisfactory. Purify inclusion bodies using preferred inclusion body purification method.
2. Refolding: Dissolve pure inclusion bodies in 6-M guanidine hydrochloride at 10 mg/mL, mixed at a ratio of three *dead* to one *alive* subunit and refolded in 100 volumes PBS. Reduce the salt concentration to 2 mM by cycles of concentration using a 10 kDa cutoff spin concentrator (Amicon) and dilution with 20-mM Tris pH 7.0.
3. Purification: Load refolding mixture onto a MonoQ 5/50 column. Wash column with 100-mM NaCl in 20-mM Tris pH 7.0. Elute mSAv-3 × His$_6$ using 20-mM Tris pH 7.0, 240-mM NaCl. Cleave off the Glu$_6$-tag overnight using human Rhinovirus 3C protease. Purify digested protein using Superdex 200 size exclusion chromatography in PBS to yield the pure anchor unit. Determine protein concentration, for example, by measuring the absorbance at 280 nm with a nanospectrophotometer and converting via the molar extinction coefficient. Concentrate the protein to 2.3-fold desired concentration and add 1.3 volumes of a solution containing 90% glycerol and 10% 10× PBS to reach a final concentration of 50% glycerol in 1× PBS.
4. **Spring unit:** Raw peptide synthesis: Raw peptides carry an N-terminal biotin group and a C-terminal azide group. The peptide should carry one cysteine and one arginine at the desired positions as fluorophore labeling sites. For a force sensor, the labeling sites should be on opposite ends of the peptide. For the control sensor, the labeling sites should be close together in order to yield a force-insensitive construct. Dissolve raw peptide in 0.1% trifluoroacetic acid in deionized water.

5. Example sequences of the force sensor unit (labeling sites in bold).
6. Force sensor: biotin-N-GG**C**GS(GPGGA)$_5$GG**K**YGGS-K (ε-N_3).
7. Control construct: biotin-N-GGGGS(GPGGA)$_5$GG**K**Y**C**GS-K(ε-N_3).
8. Purification: Purify the raw peptide via preparative C18 reverse phase HPLC (5-μm particle size, 250 × 21.2 mm, Agilent) applying a linear gradient of 0.1% trifluoroacetic acid (TFA) in deionized water (buffer A) and 0.1% TFA, 9.9% deionized water, 90% acetonitrile (buffer B) over 80 min at a flow rate of 5 mL/min. Pool the peaks of interest and lyophilize overnight.
9. Labeling: Dissolve peptide at 5 mg/mL in PBS with 400-mM Tris. Add 1.5-fold molar surplus of Alexa Fluor 555 maleimide and incubate at room temperature for 2 h. Perform HPLC purification and lyophilization. Add 1.2-fold molar surplus of Alexa 647 succinimidyl ester in the presence of 330-mM $NaHCO_3$ for 2 h at room temperature. Perform HPLC purification and lyophilization.
10. **Functional unit:** Expression of functional unit of interest in inclusion bodies using the preferred expression system: The protein of interest should feature a C-terminal unpaired cysteine residue in order to allow the conjugation to the force sensor module. Refold protein of interest with preferred method.
11. Purification: Concentrate the protein of interest using a 10 kDa cutoff spin concentrator in the presence of 25-μM Tris-(2-carboxyethyl)phosphine hydrochloride (TCEP, Pierce). Purify product via size exclusion chromatography in PBS and then concentrated to 1 mg/mL in the presence of 25-μM TCEP.
12. Labeling: Let protein of interest react with a tenfold surplus of dibenzylcyclooctyne-maleimide (DBCO) for 2 h at room temperature. Remove unreacted DBCO by gel filtration. Final concentration of the protein should be approximately 1 mg/mL.
13. **Force sensor assembly:** Click reaction: Mix labeled spring peptides with DBCO-conjugated functional unit proteins at a molar ratio of at least 5:1 in 25 vol% 2-M Tris pH 8.0. Incubate for 2 h at room temperature. Remove unreacted protein using a monomeric avidin column (Pierce). Elute protein with biotin. Remove unconjugated peptide using size exclusion chromatography on a Superdex 75 column. Determine protein concentration, for example, by measuring the absorbance at 280 nm with a nanospectrophotometer and converting via the molar extinction coefficient. Concentrate the protein to 2.3-

fold desired concentration and add 1.3 volumes of a solution containing 90% glycerol and 10% 10× PBS to reach a final concentration of 50% glycerol in 1× PBS.

14. Gel-shift assay: Use purified monovalent streptavidin in a gel shift assays to test for the successful conjugation of the labeled peptide with the DBCO-labeled protein of interest.

3.2 Creation of Antigen-Experienced T-Cell Blasts

1. Isolate T-cells from murine spleen and lymph nodes (in our case from 8 to 12 weeks old 5c.c7 aβ TCR-transgenic mice). For T-cell stimulation, priming and proliferation add antigenic HPLC-purified peptide, in our case the moth cytochrome C peptide (residue 88–102, MCC, ANERADLIAYLKQATK), to the T-cell medium. Supplement culture medium with 50 U/mL IL-2 on day 2. Expand T-cell cultures at a ratio of 1:1 on day 3 and 5 in fresh media for up to 9 days.
2. On day 6, prior to carrying out the microscopy experiments, separate living T-cells from debris and dead cells by collecting them after centrifugation from a Histopaque-1119 cushion. T-cells can be used on day 7–9 after stimulation.

3.3 Labeling of T-Cells with Fura-2

1. Add 0.5 μL of 1 mg/mL Fura-2 in DMSO to 0.5×10^6 T-cells in 500-μL T-cell medium.
2. Incubate 20 min at room temperature.
3. Add 4.5 mL of imaging buffer.
4. Centrifuge for 3 min at 350 g.
5. Remove supernatant.

3.4 Preparation of Sensor-Decorated Functionalized Supported Lipid Bilayers

1. **Lipid vesicle preparation:** Prepare lipid mixtures of 61.25 nmol base lipid and 1.25 nmol Ni-NTA-DGS (98: 2 ratio): Dissolve DPPC, POPC, and Ni-NTA-DGS in chloroform. Pipette DPPC (44.96 μg) and Ni-NTA-DGS (1.32 μg) or POPC (46.55 μg) and Ni-NTA-DGS (1.32 μg) into a glass test tube.
2. Evaporate chloroform in a nitrogen stream. Slowly turn the test tube while doing so to spread the lipids on the glass. Once all visible chloroform is evaporated, keep lipids in the nitrogen stream for at least 20 min to remove all chloroform traces.
3. Resuspend lipids in 1-mL DPBS at a temperature above the melting point of the base lipid. For POPC (melting point: −2 °C), resuspend at room temperature. For DPPC (melting point: 41 °C), dip the test tube into a warm water bath (55 °C) while mixing the tube contents with a pipette. The lipid suspension should turn slightly milky.
4. Fill the test tube with nitrogen and seal with parafilm.

5. Sonicate lipids in a water bath for 10 min. The water temperature should be above the lipid melting point, that is, room temperature for POPC, 55 °C for POPC. The sonication process is effective if the surface of the lipid suspension vibrates and droplets of liquid are occasionally ejected. This usually works best if the suspension level is a few millimeters below the water bath level. After sonication, the tube contents should be completely clear. *See* also **Notes 2** and **3**.
6. **Lipid bilayer formation:** Treat cover slips in the plasma cleaner for 10 min.
7. Remove original cover slip from an 8-well chamber and discard. Rinse the remaining plastic part with ethanol and deionized water. Dry under a nitrogen stream. *See* also **Note 4**.
8. Mix about 0.5 mL of each component of the duplicating silicone and apply to the bottom of the 8-well chamber. Attach the plasma-cleaned coverslip and wait 5 min for the silicone to harden. Check whether enough silicone was applied to isolate the wells from each other.
9. Put 150 μL of lipid vesicle solution into each well.
10. Incubate slides for 20 min at a temperature above the melting point of the lipids, that is, room temperature for POPC, 55 °C for POPC.
11. Wash away excess vesicles by rinsing each well with 30-mL DPBS.
12. Remove liquid meniscus from each well. Then, remove 330 μL from each well. This leaves 450 μL.
13. **Protein incubation:** The following protein quantities should yield concentrations in the order of 0.1 labeled force sensors per μm^2, which is suitable for single-molecule experiments.
14. Prepare protein solutions in 50-μL DPBS per well:
    (a) *Activating conditions*: 30-ng ICAM-1, 50-ng B7-1, 10-ng mSav-3 × His_6
    (b) *Scanning conditions*: 30-ng ICAM-1, 10-ng mSav-3 × His_6
    (c) *Densely labeled bilayer*: 10-ng mSav-3 × His_6
15. Add 50 μL of protein solution to each well and incubate for 75 min.
16. Wash away unbound proteins by rinsing each well with 30-mL DPBS.
17. Remove liquid meniscus and afterward 330 μL from each well.
18. Prepare force sensor solutions in 50-μL DPBS per well:
    (a) *Activating conditions*: 15-ng unlabeled sensor (IE^k-MFS_0), 170 pg labeled sensor (IE^k-MFS)

(b) *Scanning conditions*: 90 pg labeled sensor (IE^k-MFS)

(c) *Densely labeled bilayer*: 15 ng labeled sensor (IE^k-MFS)

19. Add 50 μL of force sensor solution to each well and incubate for 20 min.
20. Wash away unbound sensors by rinsing each well with 30-mL DPBS.
21. Remove liquid meniscus and finally 330 μL from each well.
22. **Imaging buffer exchange:** Exchange the DPBS in a well for imaging buffer (HBSS +2% FCS) just before using the well for experiments. While bilayers are stable for hours in DPBS, they degrade typically within an hour in imaging buffer.
23. Remove 300 μL of liquid from the well.
24. Add 300 μL of imaging buffer to the well.
25. Repeat preceding steps four times.

3.5 Microscopy

1. **Ensure reactivity of cells towards the sensor construct:** For instance, perform calcium flux measurements using the Fura-2-labeled T-cells [15]
2. **Setup for single-molecule FRET measurements:** Use TIRF configuration to minimize cellular background and to maximize signal-to-noise ratio.
3. If using an EM-CCD camera, enable the EM-gain.
4. **Stroboscopic alternating excitation (ALEX):** See Fig. 2b for a visualization of the alternating excitation sequence. Recorded sample images are displayed in Fig. 3a.
5. Register cell-occupied region by exciting Fura-2-loaded T-cells with 405-nm light for a time t_{ill} (typically 5–10 ms) while triggering the camera. The emission around 510 nm is captured in the donor emission channel.
6. With excitation light sources turned off, wait for a time t_{r} for the camera to read out the chip. The recorded image can later be used for image segmentation. On EM-CCD cameras, t_{r} can be minimized (typically 5 ms) by reducing the number of readout pixel lines.
7. Excite the donor fluorophores using the 532-nm laser for t_{ill} while triggering the camera.
8. Read out donor (Im_{DD}) and FRET emission (Im_{DA}) image by waiting for t_{r}.
9. Excite acceptor fluorophores using the 640-nm laser for t_{ill} while triggering the camera.

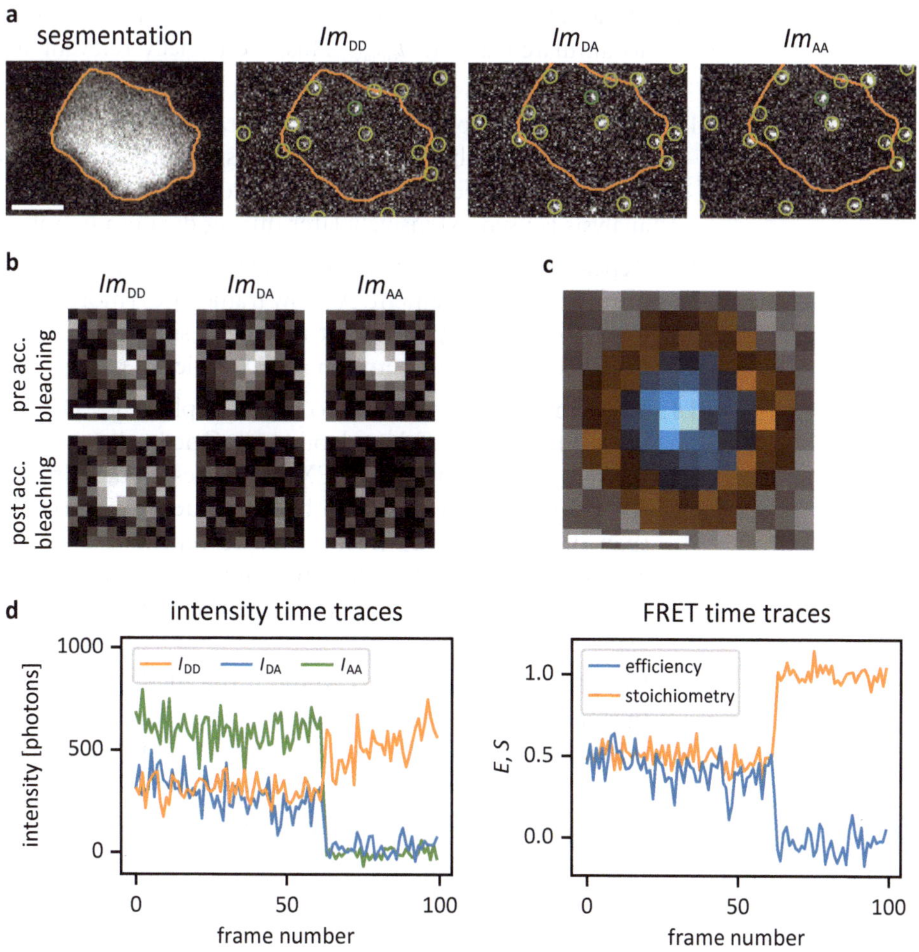

Fig. 3 Single-molecule image analysis. (**a**) Sample images. The segmentation image is used to determine cell contours (orange line) via adaptive thresholding. Analysis is restricted to molecules within this contour. Sensors are identified in donor (Im_{DD}), FRET (Im_{DA}), and acceptor (Im_{AA}) images by means of a single-molecule localization algorithm. All molecules marked by a circle could be tracked across at least four frames and were subsequently analyzed. Data from the sensor highlighted by the green circle is shown in the following panels. Scale bar: 5 μm. (**b**) Crops of the region around a single molecule, which was observed over 100 consecutive frames in total. For display, we selected here representative images before (top row) and after (bottom row) photobleaching of the acceptor. Scale bar: 1 μm. (**c**) To determine the fluorescence intensities (I_{DD}, I_{DA}, and I_{AA}), pixel intensities around the molecule's center (blue) are summed up. Local background is computed from pixels in a ring around this circle (orange) and subtracted. The inner circle needs to be chosen large enough to contain the molecule's point spread function. Scale bar: 1 μm. (**d**) Time traces of the fluorescence intensities calculated as described (left) and the derived FRET quantities (right). Acceptor photobleaching is visible as a sudden drop in I_{DA}, I_{AA}, and E and a concomitant increase in I_{DD} and S

10. Wait for a time t_{delay} ($\geq t_r$) to read out the acceptor emission image (Im_{AA}) and to adjust the frame rate. As the length of single-molecule trajectories is limited by fluorophore (mostly, acceptor) photobleaching, reducing the frame rate allows for prolonged observation periods. Choose $t_{delay} = t_r$ to record as fast as possible.

11. Optionally, read out the camera chip to remove any charges accumulated during t_{delay}. This is especially relevant for long delays.
12. Repeat the five preceding steps several (typically, hundreds of) times. The number of ALEX cycles should be high enough to warrant photobleaching of all acceptors, as step-wise bleaching analysis is used extensively later during data evaluation.
13. Repeat steps one and two to visualize the movement of the cell during the measurement. At slow frame rates (more specifically, long observation periods), it may be desirable to record additional cell images in between ALEX cycles.
14. **Experiments:** Add T-cells to each sample (e.g., DPPC-based lipid bilayer, POPC-based bilayer). Once cells attach to the bilayer, start recording ALEX sequences as described above. After 20 min—this time can be varied depending on the time span of interest—add fresh cells and repeat. We recommend recording at least 100 videos per sample. For quantitative data analysis, it is required to perform the following calibration experiments.
15. Samples without cells: To establish the FRET efficiency distributions of the sensors at zero force, ALEX sequences of bilayers without cells need to be recorded. We recommend taking a minimum of 30 videos per sample.
16. Fiducial markers: Registration of donor and acceptor emission channel coordinates requires images of randomly distributed fiducial markers, which are visible in both channels. Record multiple images, moving the markers in between acquisitions such that the whole field of view is covered.
17. Densely labeled bilayer: For flatfield correction, acquire the excitation laser profiles as follows: Record emission upon acceptor excitation, subsequently bleach all acceptor fluorophores, and finally record emission upon donor excitation. Repeat at least five times.
18. Special samples for quantification of FRET correction factors: All correction factors [27] can be calculated from single-molecule ALEX traces of samples without cells without any need for additional measurements. Traces with bleached acceptors yield the donor leakage and traces with bleached donors yield direct acceptor excitation; detection efficiency is found by examining intensity changes upon acceptor bleaching of collapsed sensors; excitation efficiency is calculated from the apparent stoichiometry of a single-molecule population of sensors. However, if desired, the analysis software also supports computation of correction factors from ALEX traces of donor-only samples, acceptor-only samples, and samples featuring at least two discrete FRET states (e.g., Holliday junctions [28]), respectively.

3.6 Data Analysis

The analysis pipeline is organized as a series of *Jupyter notebooks* containing code cells including extensive comments, in which well-documented analysis routines are called. It can be accessed at https://github.com/schuetzgroup/fret-analysis. Documentation can be displayed by typing a method name followed by a question mark? after a method name, e.g., tr.track?.

Execute the code cells in sequence. Key functionality is described below. Note that we have previously published an extensive guide on how to analyze single-molecule FRET data [29], notably featuring screenshots and explanations of each code cell in a supplementary document.

A flow chart summarizing the analysis is shown in Fig. 4.

1. **Setup:** Use git to create a new folder for analysis.
2. `git clone https://github.com/schuetzgroup/fret-analysis <target_folder>`
3. Replace <*target_folder*> by the desired folder name. It must not exist beforehand.
4. Copy Jupyter notebooks to the newly created folder. If starting from scratch, copy *01. Tracking.ipynb*, *02. Analysis.ipynb*, and *03. Plots.ipynb* from the *notebooks* subfolder. It may sometimes be convenient to copy notebooks from a previous analysis, for instance, if only slight adaptions are expected.
5. Start *Jupyter Lab* from the Anaconda installation, wait for the application to load in a web browser window, then navigate to the analysis folder via the left pane.
6. **Single-molecule localization, tracking, and intensity determination:** Open the *01. Tracking.ipynb* notebook in the Jupyter Lab window.
7. In the appropriate code cell, describe data by setting the data directory and excitation sequence. Identify datasets by specifying *regular expressions* matching corresponding raw data file names. Load data for image registration and flatfield correction.
8. If using a single camera to capture both donor and acceptor emission, use the provided interface to select the respective regions on the camera chip.
9. Choose a localization algorithm and appropriate parameters to detect fiducial markers for image registration. Do this for both donor and acceptor emission channels. Then, perform the registration (i.e., mapping coordinates of one channel to the other and vice versa).
10. Select a localization algorithm and suitable parameters for detection of sensor molecules separately in (a) the sum of

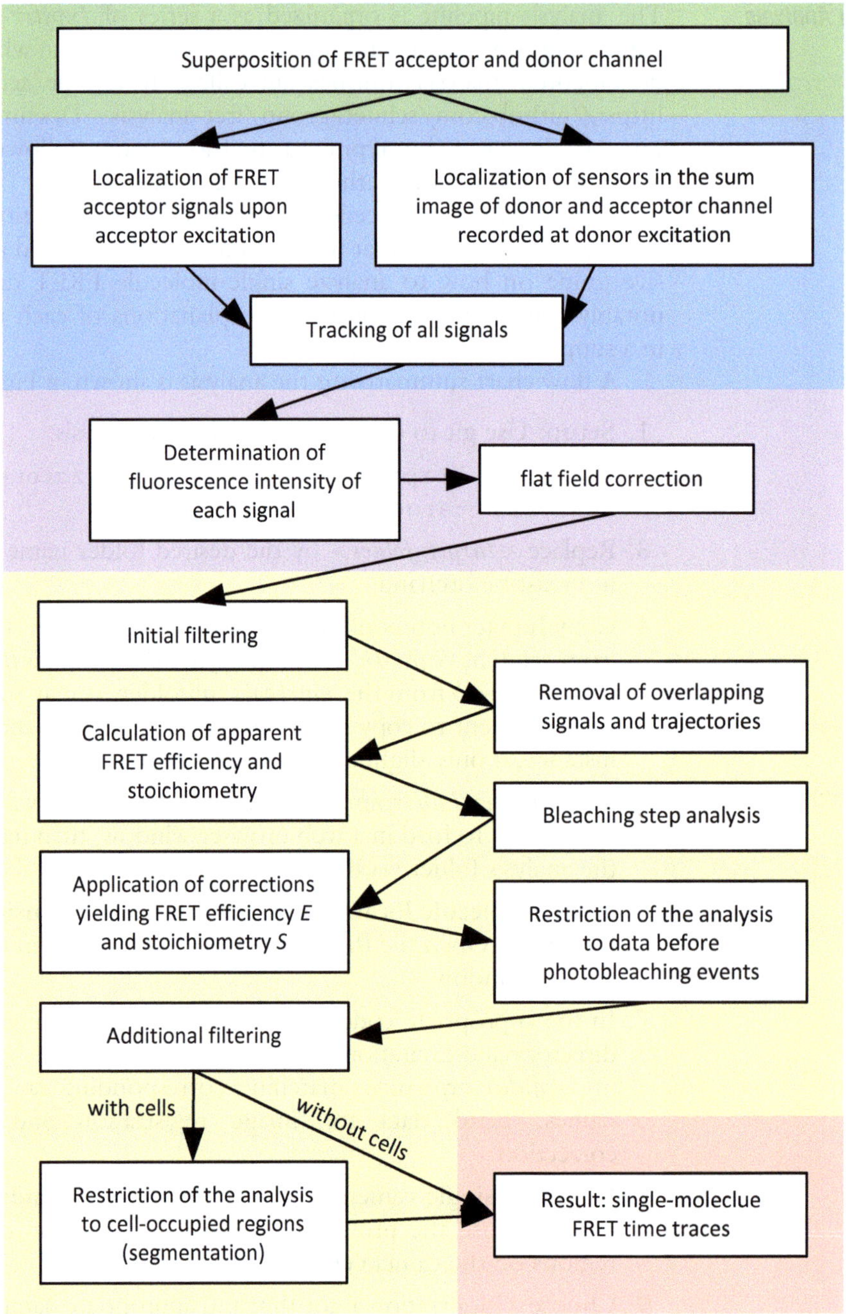

Fig. 4 Analysis pipeline. Summary of the analysis steps to extract single-molecule FRET time traces from raw image data

donor and acceptor emission images upon donor excitation and (b) the acceptor emission image upon acceptor excitation. *See* also **Note 5**. Consequently, execute the algorithm. This can take a while.

11. Link molecule localizations into trajectories using a tracking algorithm while also determining fluorescence intensities. The parameters the algorithms can be estimated by visual inspection of raw data in an image viewer. *See* Fig. 3 for an illustrative example.
12. Extract additional image data, such as images for segmentation and flatfield correction calculation.
13. Save data to disk. This creates several files with the same, user-defined prefix (`"tracking"` by default).
14. **smFRET analysis and filtering:** Open the *02. Analysis.ipynb* notebook in the Jupyter Lab window.
15. Load data created previously.
16. Optionally, select only molecules residing in sufficiently illuminated regions of the camera chip. This is done by thresholding the illumination profile recorded for flatfield correction.
17. Optionally, perform additional initial filtering. For instance, it may be desired to discard tracks that are very close to other tracks, as fluorescence intensity determination is not reliable if point spread functions overlap.
18. Apply flatfield correction.
19. Compute apparent FRET efficiencies and stoichiometries for each single-molecule localization. These have not been corrected for experimental deficiencies, such as donor leakage into the acceptor channel, yet.
20. Choose appropriate parameters for bleaching step detection in (a) the sum of donor and acceptor emission intensity upon donor excitation and (b) the acceptor emission intensity upon acceptor excitation. For this, it is recommended to first find the right orders of magnitude of the `penalty` parameter and then fine-tune. Subsequently, execute the step detection algorithm.
21. Optionally, dispose of tracks that do not exhibit the desired bleaching step behavior. For instance, remove all tracks where signals do not bleach in a single step as such behavior may reflect the presence of aggregates.
22. Compute FRET correction factors according to [27]. Donor emission leakage into the acceptor channel can be derived either from parts of molecules' time traces where the acceptor was bleached or from a donor-only sample. Direct acceptor excitation by the donor excitation light source may be determined from parts of molecules' time traces where the donor was bleached or from an acceptor-only sample. Differences in detection efficiencies and quantum yields for donor and acceptor can be accounted for by analyzing intensity changes upon acceptor bleaching in traces of high FRET efficiency (e.g.,

collapsed sensors from a sample without cells) [30] or a sample with probes featuring two distinct states (e.g., Holliday junctions). Quantification of discrepant excitation efficiencies may be performed by selecting and evaluating a fraction of force sensor data with a 1:1 acceptor-to-donor stoichiometry [31] or, again, by analyzing a sample featuring two distinct states.

23. After calculating all four factors, apply them to the data sets.
24. From each sensor time trace, select only data from before the first bleaching event.
25. Optionally, apply additional filters. For instance, accept only traces where stoichiometries lie within a specified range.
26. Optionally, perform image segmentation using the Fura-2-emission images of cells. For this, select suitable parameters for adaptive thresholding. Subsequently, reject all signals from sensors outside cell boundaries.
27. Save data to disk. This creates a file with a user-defined prefix (`"filtered"` by default).
28. **Visualization of results and further analysis:** Open the *03. Plots.ipynb* notebook in the Jupyter Lab window.
29. Create FRET efficiency-vs.-stoichiometry plots and FRET efficiency histograms. Exemplary results are shown in Fig. 5a, b.
30. Use the sensor calibration to convert FRET efficiencies to forces (Fig. 5c).
31. Perform further analysis such as diffusion analysis and spatio-temporal analysis of time traces [15] or hypothesis testing to find differences between datasets; *see* also **Note 6**.

4 Notes

1. Some batches of fetal calf serum exhibit fluorescence, leading to substantial background in the acceptor emission channel, particularly with POPC-based bilayers. It is recommended to test different batches to minimize the problem.
2. Sonication of the lipid suspension can be tricky. Effectiveness varies with the ultrasound water bath, test tubes, lipid suspension volume, and bath water level. The ultrasound bath and test tubes presented here have been found to work well if the bath is filled to roughly 2 cm below the rim. Larger amounts of vesicles than described here can be produced by increasing the quantity of lipids suspended in 1-mL DPBS and diluting proportionally prior to bilayer formation.
3. Lipid vesicles can be kept for several weeks if stored under inert gas such as N_2 or Ar. We suggest storing POPC-based vesicles

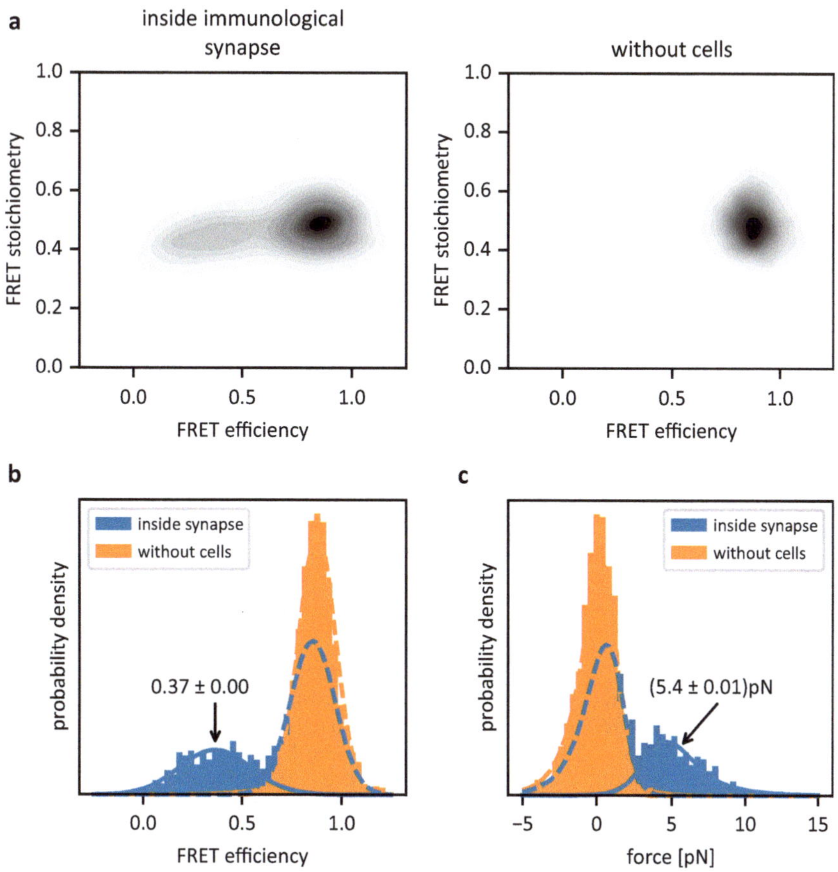

Fig. 5 Exemplary results. (**a**) FRET efficiency versus stoichiometry plots for sensors within the immunological synapse (left) and for a sample containing sensors on an SLB without cells. Stoichiometry is around 0.5, indicating a 1:1 donor-to-acceptor ratio. A shift toward lower FRET efficiencies can be clearly seen inside the immunological synapse, revealing the presence of forces. Density of data points is indicated by the shade from low (white) to high (black). (**b**) FRET efficiency histograms of the datasets from (**a**). Lines indicate the results of a two-component Gaussian mixture model fit for each dataset. (**c**) Conversion of FRET efficiencies from (**b**) to forces using the sensor calibration (*see* also Fig. 1b). The displayed data are part of a larger, previously published dataset [15]

at 4 °C and DPPC-based vesicles at room temperature to prevent precipitation.

4. The plastic parts the 8-well chambers can be stored and reused. After the experiment, remove the coverslip and duplicating silicone, rinse with ethanol and deionized water, and dry under nitrogen. Store in a clean environment.

5. Using the sum of donor and acceptor emission images upon donor excitation for locating sensor molecules allows for detection irrespective of their FRET efficiency.

6. Data points within single-molecule time traces are correlated and thus not independent. This requires extra care when performing statistical hypothesis tests [32].

Acknowledgments

This work was supported by the Austrian Science Fund (FWF) projects P32307-B and by the Vienna Science and Technology Fund (WWTF) 10.47379/LS13030.

References

1. Yusko EC, Asbury CL (2014) Force is a signal that cells cannot ignore. Mol Biol Cell 25: 3717–3725. https://doi.org/10.1091/mbc.e13-12-0707
2. Chen Y, Ju L, Rushdi M et al (2017) Receptor-mediated cell mechanosensing. Mol Biol Cell 28:3134–3155. https://doi.org/10.1091/mbc.e17-04-0228
3. Huang J, Zarnitsyna VI, Liu B et al (2010) The kinetics of two-dimensional TCR and pMHC interactions determine t-cell responsiveness. Nature 464:932–936. https://doi.org/10.1038/nature08944
4. Huppa JB, Axmann M, Mörtelmaier MA et al (2010) TCR–peptide–MHC interactions in situ show accelerated kinetics and increased affinity. Nature 463:963–967. https://doi.org/10.1038/nature08746
5. Klotzsch E, Schütz GJ (2013) Improved ligand discrimination by force-induced unbinding of the t cell receptor from peptide-MHC. Biophys J 104:1670–1675. https://doi.org/10.1016/j.bpj.2013.03.023
6. Schütz GJ, Huppa JB (2019) How drag sharpens a t cell's view on antigen. Proc Natl Acad Sci 116:16669–16671. https://doi.org/10.1073/pnas.1911389116
7. Thomas WE, Vogel V, Sokurenko E (2008) Biophysics of catch bonds. Annu Rev Biophys 37:399–416. https://doi.org/10.1146/annurev.biophys.37.032807.125804
8. Feng Y, Brazin KN, Kobayashi E et al (2017) Mechanosensing drives acuity of αβT-cell recognition. Proc Natl Acad Sci 114:E8204–E8213. https://doi.org/10.1073/pnas.1703559114
9. Hu KH, Butte MJ (2016) T cell activation requires force generation. J Cell Biol jcb.201511053. https://doi.org/10.1083/jcb.201511053
10. Liu B, Chen W, Evavold BD, Zhu C (2014) Accumulation of dynamic catch bonds between TCR and agonist peptide-MHC triggers t cell signaling. Cell 157:357–368. https://doi.org/10.1016/j.cell.2014.02.053
11. Liu Y, Blanchfield L, Ma VP-Y et al (2016) DNA-based nanoparticle tension sensors reveal that t-cell receptors transmit defined pN forces to their antigens for enhanced fidelity. Proc Natl Acad Sci USA 201600163. https://doi.org/10.1073/pnas.1600163113
12. Grakoui A, Bromley SK, Sumen C et al (1999) The immunological synapse: a molecular machine controlling t cell activation. Science 285:221–227. https://doi.org/10.1126/science.285.5425.221
13. Aramesh M, Mergenthal S, Issler M et al (2021) Functionalized bead assay to measure three-dimensional traction forces during t-cell activation. Nano Lett 21:507–514. https://doi.org/10.1021/acs.nanolett.0c03964
14. Fölser M, Motsch V, Platzer R et al (2021) A multimodal platform for simultaneous t-cell imaging, defined activation, and mechanobiological characterization. Cells 10. https://doi.org/10.3390/cells10020235
15. Göhring J, Kellner F, Schrangl L et al (2021) Temporal analysis of t-cell receptor-imposed forces via quantitative single molecule FRET measurements. Nat Commun 12:2502. https://doi.org/10.1038/s41467-021-22775-z
16. Jin W, Tamzalit F, Chaudhuri PK et al (2019) T cell activation and immune synapse organization respond to the microscale mechanics of structured surfaces. Proc Natl Acad Sci 116: 19835–19840. https://doi.org/10.1073/pnas.1906986116
17. Plotnikov SV, Sabass B, Schwarz US, Waterman CM (2014) Chapter 20: High-resolution traction force microscopy. In: Waters JC, Wittman T (eds) Quantitative imaging in cell biology. Academic, pp 367–394
18. Saez A, Ghibaudo M, Buguin A et al (2007) Rigidity-driven growth and migration of epithelial cells on microstructured anisotropic substrates. Proc Natl Acad Sci 104:8281–8286. https://doi.org/10.1073/pnas.0702259104
19. Bashour KT, Gondarenko A, Chen H et al (2014) CD28 and CD3 have complementary roles in t-cell traction forces. Proc Natl Acad Sci 111:2241–2246. https://doi.org/10.1073/pnas.1315606111

20. Colin-York H, Javanmardi Y, Skamrahl M et al (2019) Cytoskeletal control of antigen-dependent t cell activation. Cell Rep 26: 3369–3379.e5. https://doi.org/10.1016/j.celrep.2019.02.074
21. Becker N, Oroudjev E, Mutz S et al (2003) Molecular nanosprings in spider capture-silk threads. Nat Mater 2:278–283. https://doi.org/10.1038/nmat858
22. Grashoff C, Hoffman BD, Brenner MD et al (2010) Measuring mechanical tension across vinculin reveals regulation of focal adhesion dynamics. Nature 466:263–266. https://doi.org/10.1038/nature09198
23. Zhang Y, Ge C, Zhu C, Salaita K (2014) DNA-based digital tension probes reveal integrin forces during early cell adhesion. Nat Commun 5:5167. https://doi.org/10.1038/ncomms6167
24. Ma VP-Y, Liu Y, Blanchfield L et al (2016) Ratiometric tension probes for mapping receptor forces and clustering at intermembrane junctions. Nano Lett 16:4552–4559. https://doi.org/10.1021/acs.nanolett.6b01817
25. Brenner MD, Zhou R, Conway DE et al (2016) Spider silk peptide is a compact, linear nanospring ideal for intracellular tension sensing. Nano Lett 16:2096–2102. https://doi.org/10.1021/acs.nanolett.6b00305
26. conda-forge Contributors (2023) Miniforge. https://github.com/conda-forge/miniforge. Accessed 9 May 2023
27. Hellenkamp B, Schmid S, Doroshenko O et al (2018) Precision and accuracy of single-molecule FRET measurements—a multi-laboratory benchmark study. Nat Methods 15:669. https://doi.org/10.1038/s41592-018-0085-0
28. Schrangl L, Göhring J, Schütz GJ (2018) Kinetic analysis of single molecule FRET transitions without trajectories. J Chem Phys 148: 123328. https://doi.org/10.1063/1.5006038
29. Schrangl L, Göhring J, Kellner F et al (2021) Automated two-dimensional spatiotemporal analysis of mobile single-molecule FRET probes. J Vis Exp:e63124. https://doi.org/10.3791/63124
30. McCann JJ, Choi UB, Zheng L et al (2010) Optimizing methods to recover absolute FRET efficiency from immobilized single molecules. Biophys J 99:961–970. https://doi.org/10.1016/j.bpj.2010.04.063
31. Lee NK, Kapanidis AN, Wang Y et al (2005) Accurate FRET measurements within single diffusing biomolecules using alternating-laser excitation. Biophys J 88:2939–2953. https://doi.org/10.1529/biophysj.104.054114
32. Schneider MC, Schütz GJ (2022) Don't be fooled by randomness: valid p-values for single molecule microscopy. Front Bioinf 2:811053. https://doi.org/10.3389/fbinf.2022.811053

Chapter 12

A Pipeline for Dynamic Analysis of Mitochondrial Content in Developing T Cells: Bridging the Gap Between High-Throughput Flow Cytometry and Single-Cell Microscopy Analysis

Vaibhav Dhyani, Anchi S. Chann, Lopamudra Giri, Sarah M. Russell, and Mirren Charnley

Abstract

Analyzing the dynamics of mitochondrial content in developing T cells is crucial for understanding the metabolic state during T cell development. However, monitoring mitochondrial content in real-time needs a balance of cell viability and image resolution. In this chapter, we present experimental protocols for measuring mitochondrial content in developing T cells using three modalities: bulk analysis via flow cytometry, volumetric imaging in laser scanning confocal microscopy, and dynamic live-cell monitoring in spinning disc confocal microscopy. Next, we provide an image segmentation and centroid tracking-based analysis pipeline for automated quantification of a large number of microscopy images. These protocols together offer comprehensive approaches to investigate mitochondrial dynamics in developing T cells, enabling a deeper understanding of their metabolic processes.

Key words DN3, Mitochondria, Density-based spatial clustering of applications with noise (DBSCAN), Image segmentation, 3D imaging, MitoView Green

1 Introduction

Mitochondria, the organelles responsible for cellular energy production, exist in a dynamic continuum of states, and their interplay with cellular processes is critical for T cell development and differentiation [1–5]. Notably, mitochondrial respiration has been observed to peak during the double-negative-3 (DN3) stage [6]. This stage holds particular significance in T cell development, as it marks a crucial phase where T cells undergo genetic reconfiguration to form a suitable β-chain of the T cell receptor (TCR) [7], underscoring the importance of mitochondrial respiration.

Christoph Wülfing and Robert F. Murphy (eds.), *Imaging Cell Signaling*, Methods in Molecular Biology, vol. 2800,
https://doi.org/10.1007/978-1-0716-3834-7_12,

Table 1
Comparison of flow cytometry, LSCM, and SDCM for live cell imaging: Key characteristics and capabilities, strength to weakness color-coded from green to orange and yellow

Technique	Flow cytometry	LSCM	SDCM
Principle	Bulk analysis	High-resolution, 3D imaging, live cell imaging	High-speed, 3D imaging, live cell imaging
Sample throughput	High	Low to moderate	Moderate
Spatial resolution	Low	High	Moderate
Optical sectioning (3D imaging)	No	Excellent	Good
Image acquisition speed	Fast	Moderate	Fast
Phototoxicity	Low	Moderate to high	Low
Image quality	NA (except flow cytometry imaging)	Excellent	Good
Live cell time lapse	No	Limited	Excellent

To gain deeper insights into the role of cellular metabolism in determining T cell fate during β-chain selection, it is essential to analyze mitochondrial dynamics, with a particular focus on the DN3 stage [4, 6]. While conventional flow cytometry offers a valuable assessment of mitochondrial content across T cell populations, it does not capture the morphological intricacies of the mitochondrial network and does not allow for longitudinal measurements of single cells. Microscopy presents a promising avenue for addressing these limitations but poses different challenges. Imaging techniques like laser scanning confocal microscopy (LSCM) provide valuable insights into the volumetric characteristics of mitochondrial structures, but the high laser power and longer exposure times can lead to photobleaching and phototoxicity, impacting cell viability. As a result, prolonged imaging sessions for multiple days are not feasible with this technique. While live cell imaging for shorter durations, typically ranging from minutes to a few hours, is possible with careful optimization of imaging parameters, long-term imaging studies necessitate alternative approaches. Spinning disc confocal microscopy (SDCM) is one such alternative that allows longer imaging durations but at the expense of image resolution (*see* Table 1 and **Note 1**) [8]. Moreover, the choice of a microscope objective is pivotal, as higher magnification necessitates more precise scanning, potentially leading to slower imaging processes [9–11]. Therefore, optimal image quality and the preservation of cell viability must be balanced to achieve reliable microscopy-based observations [8].

To obtain high-throughput information on mitochondrial content from microscopy images, unbiased quantification of images is required, ideally involving automatic image processing, segmentation, and quantification. Another challenge in dynamically monitoring developing T cells is their mobility, which makes it challenging to track the spatiotemporal properties of the cells. This necessitates image segmentation at each time point and the tracking of individual cells over time.

In this chapter, we present a comprehensive protocol for staining and quantifying mitochondria in T cells using flow cytometry and imaging, including dynamic monitoring of single cells. Our experimental protocol is based on MitoView Green, a potential-independent dye that accumulates in the mitochondria and interacts with the mitochondrial matrix via hydrophobic interactions (*see* **Note 2**). The fluorescence intensity of MitoView Green is thus directly proportional to their mitochondrial mass or content [12, 13]. We have combined mitochondrial staining with antibody staining using fluorophore-conjugated antibodies to label T cell surface (CD4/CD8/CD44/CD25/CD28) so that the mitochondrial content of distinct populations of T cells can be analyzed [14]. Our experimental pipeline consists of three distinct modalities, each of which contributes to a comprehensive analysis of the mitochondrial content. First, we analyze cells in bulk using flow cytometry [15, 16]. Then, we perform endpoint LSCM image capturing of the sorted cells. We have also developed a methodology for live cell imaging of developing T cells on a spinning disk microscope. This technique allows us to track the dynamics of mitochondrial content changes over time, providing valuable context about behavior during T cell development.

We then introduce an image analysis pipeline, including the codes for computation that enables automated qualification of mitochondria from a large number of images and efficient tracking of the spatiotemporal behavior of mitochondria in developing T cells. Our image segmentation framework is based on density-based spatial clustering of applications with noise (DBSCAN) [17, 18]. In this approach, the problem of image segmentation is regarded as the problem of partitioning pixels into different clusters according to their spatial relationships. DBSCAN does not require the user to specify the number of clusters (or the number of cells) in advance. This is a region-based method that requires the pixels in each segmented region to be spatially connected. DBSCAN has the ability to classify noisy or outlier pixels as noise rather than assigning them to any specific cluster. This helps in reducing the impact of noise on the segmentation results, leading to the more accurate delineation of cells present in any image (*see* **Notes 3** and **4**).

Our staining and imaging protocols can facilitate the investigation of mitochondrial dynamics in developing T cells and aid in the understanding of cellular processes involved in cell differentiation and proliferation. While we focus on the DN3 stage of T cell development, we conjecture that these protocols can also be implemented to study mitochondria in other cell types.

2 Materials and Software

2.1 Reagents

1. C57BL/6 mouse E14.5 fetal liver
2. OP9-DL1 stromal cells
3. FACS buffer: PBS supplemented with 2% FBS

2.2 Cell Culture Media

1. OP9-DL1 culture media: Minimal Essential Media alpha (MEMα, SAFC®) supplemented with 20% FBS, 1 mM glutaMAX, and 100 ng/mL penicillin/streptomycin.
2. Stem cell media: Anne Kelso media supplemented with 10% FBS, 50 μM β-mercaptoethanol, and 100 ng/mL penicillin/streptomycin and 1 mM L-asparagine.
3. Fetal liver resting media: stem cell media supplemented with 20% FBS, 10 ng/mL SCF, 2 ng/mL mouse IL-6 (PeproTech), and 2 ng/mL mouse IL-3 (PeproTech) (supplement with cytokines on the day of use).
4. T cell media: MEMα supplemented with 10% FBS, 1 mM glutaMAX, 10-mM HEPES, 1 nM sodium pyruvate, 100 ng/mL penicillin/streptomycin, 50 μM β-mercaptoethanol, 1 ng/mL mouse IL-7 (PeproTech), and 5 ng/mL mouse FMS-like tyrosine kinase 3 (Flt3) (PeproTech) (supplement with cytokines on the day of use).

2.3 Stains/Antibodies

1. BV421 CD4 antibody (Biolegend, Catalog number: 100543, Clone: RM4-5)
2. BV711 CD8 antibody (BD Biosciences, Catalog number: 563046, Clone: 53-6.7)
3. BV510 CD44 antibody (BD Biosciences, Catalog number: 563114, Clone: IM7)
4. BV786 CD25 antibody (BD Biosciences, Catalog number: 564023, Clone: PC61)
5. Pe-Cy7 CD28 antibody (Biolegend, Catalog number: 122014, Clone: E18)
6. MitoView Green 490/523 nm (Biotium, Catalog number: 70054)
7. SiR-DNA 652/674 nm (Cytoskeleton, Inc., Catalog number: SC007)

2.4 Cell Culture Plasticware

1. Glass bottom dish 35 mm (Ibidi)
2. Eight-well optical bottom chamber slides (Ibidi)
3. PDMS cell paddocks (grid size 250 μm × 250 μm) (prepared inhouse via spin coating the PDMS onto a silicon master. They can alternatively be purchased from Microsurfaces Pty Ltd., Melbourne, Australia)

2.5 Equipment

1. Flow cytometer (BD Biosciences, BD FACSAria™ III)
2. Laser scanning confocal microscope (Olympus Life Science, FV3000)
3. Spinning disc confocal microscope (Olympus IX71), with CSU 10 scanner and ANDOR iXon3 EMCCD camera

2.6 Software and Package Requirement

1. MetaMorph version 7.10.3
2. FlowJo™ v10.8.1
3. MATLAB R2023a with Image Processing Toolbox installed

3 Methods

3.1 Coculture of HSCs and OP9-DL1 Stromal Cells

1. Seed 1–2 × 10^5 OP9-DL1 cells suspended in 2 mL of OP9-DL1 culture media per well in a 6-well plate.
2. Thaw a frozen vial (containing 2–3 × 10^6 cells) of hematopoietic stem cells (HSCs) derived from C57BL/6 mouse E14.5 fetal liver in the fetal liver resting media in a separate culture plate.
3. After 24 h, harvest the HSCs by pipetting and centrifuge at 2042× *g* for 4 min to obtain a cell pellet.
4. Resuspend the HSC pellet in T cell media.
5. Aspirate the OP9-DL1 culture media from the 6-well plate culture of OP9-DL1 cells.
6. Add 2–4 × 10^5 HSCs in each well of the 6-well plate, pre-seeded with OP9-DL1 cells.
7. Incubate the co-culture at 37 °C and 10% CO_2. To obtain a sufficient number of DN3 populations, T cells should be expanded for two-three passages, where expansion is performed when cells are confluent.

3.2 Harvesting T Cells and Expanding the Co-culture

1. After the co-culture of HSCs with OP9-DL1 stromal cells reaches confluence, harvest the cells from the 6-well plates using forceful pipetting (Fig. 1a).
2. Using a sterile pipette, carefully transfer the cell suspension from each well to a sterile 50 mL centrifuge tube.
3. Pulse-spin the tube for 2 s at 2042× *g* (Fig. 1a).

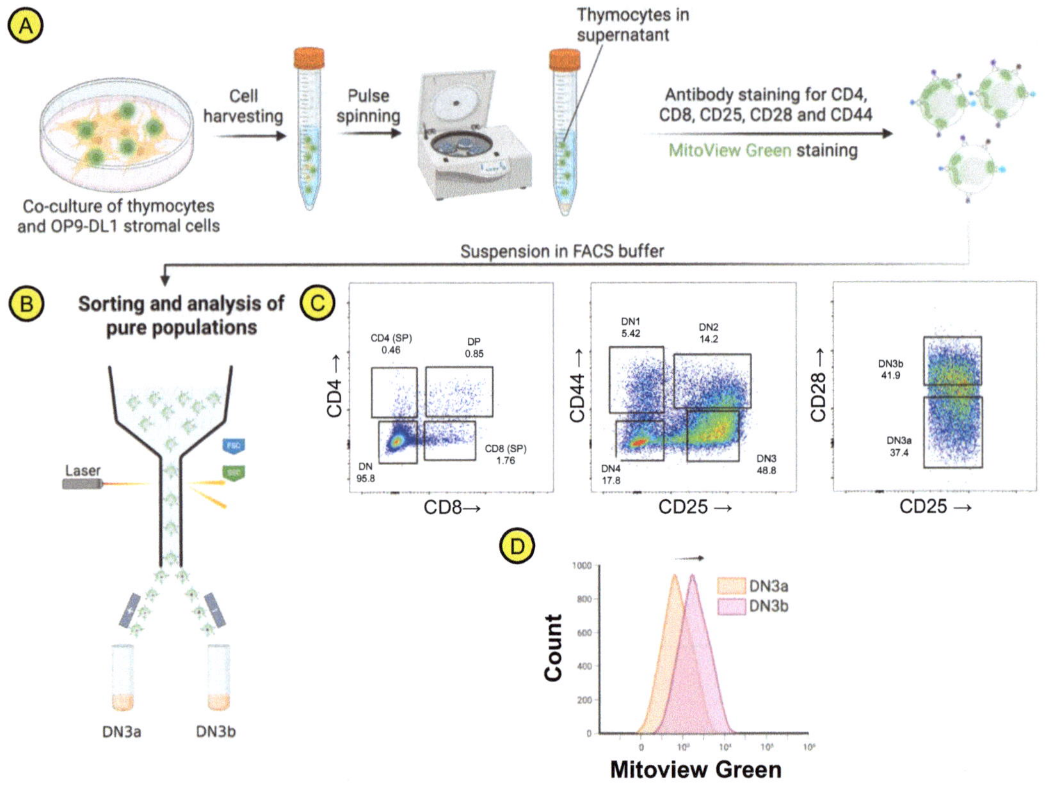

Fig. 1 Workflow for bulk analysis of mitochondria in T cells using flow cytometry. (**a**) T cells are harvested by forced pipetting and pulse spinning after co-culturing of HSCs on OP9-DL1. The T cell pellet obtained after centrifuging the supernatant is resuspended in FACS buffer containing CD4, CD8, CD44, CD25, and CD28 antibodies, along with MitoView Green. (**b**) Pure populations of DN3a and DN3b cells are obtained through cell sorting. (**c**) Gating strategy to obtain pure populations of DN3a and DN3b cells. (**d**) The distribution of the green emission from each cell analyzed from flow cytometry is presented as a histogram

4. Carefully transfer the supernatant to a new sterile 50 mL microcentrifuge tube. The supernatant contains T cells, while the OP9-DL1 cells will have settled at the bottom of the original tube (Fig. 1a).
5. Centrifuge the tube containing the T cells at 2042× *g* for 4 min to pellet the cells.
6. To passage the culture, resuspend the T cells in T cell media.
7. Seed the resuspended T cells onto a fresh 6-well plate already seeded with 1–2 × 10^5 OP9-DL1 cells a day prior (same as **steps 5** and **6** in Subheading 3.1).

3.3 Staining with FACS Antibodies and MitoView Green

1. Prepare 1 mL of antibody and MitoView Green staining solution in FACS buffer as per the dilutions mentioned in Table 2.

Table 2
Dilutions of antibody stains and fluoroprobes used in the experimental protocol

S.No.	Stain	Dilution
1	BV421 CD4 antibody	1/500
2	BV711 CD8 antibody	1/1000
3	BV510 CD44 antibody	1/500
4	BV786 CD25 antibody	1/1000
5	Pe-Cy7 CD28 antibody	1/200
6	MitoView Green	200 nM

2. Resuspend the cell pellet obtained after harvesting the T cells in the staining solution prepared in the above step and transfer the cell suspension to a sterile 15 mL centrifuge tube.
3. Incubate the cells at 4 °C for 45–60 min, protected from light.
4. After incubation, wash the cells by adding 4 mL excess FACS buffer in the centrifuge tube and centrifuge the cells at 2042× *g* for 4 min.
5. Discard the supernatant to remove excess dye.
6. Resuspend the cells in 1 mL T cell media and transfer to a sterile round bottom polystyrene test tube after filtering through a 35 μm nylon mesh strainer snap cap.
7. The cell suspension is now ready for cell sorting and analysis (Fig. 1b) (*see* **Note 5**).

3.4 High-Throughput Analysis with Flow Cytometry

1. Obtain pure populations of DN3a and DN3b cells using the gating strategy shown in Fig. 1c. Pure populations are collected in FACS tubes containing 500 μL FBS.
2. Prior to sorting the cells, record the fluorescent intensities of the cells in both the DN3a and DN3b categories. This will allow for a comparison of the mitochondrial content between the two populations.
3. Prepare a histogram plot of the fluorescence intensities for both DN3a and DN3b populations by using flow cytometry analysis software (FlowJo).
4. Compare the histogram plots of the mitochondrial content in individual cells for DN3a and DN3b cells. The plot with higher fluorescence intensity represents the population with greater mitochondrial content (Fig. 1d).

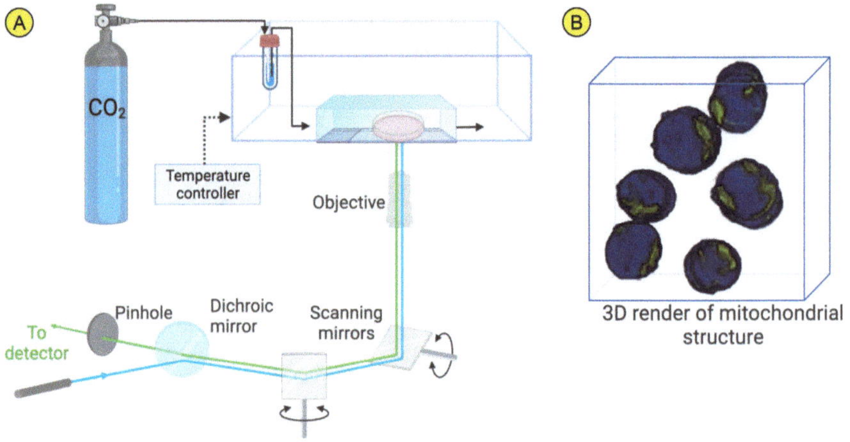

Fig. 2 Snap-shot assay with LSCM. (**a**) The pure populations of DN3a and DN3b cells are imaged using a LSCM. (**b**) 3D imaging is performed by capturing serial stacks across the cell height, enabling the construction of 3D renders

3.5 Snap-Shot Assay with LSCM

1. Centrifuge the pure populations of DN3a and DN3b cells at 2042× *g* for 4 min. Remove the supernatant and carefully resuspend each cell population in 2 mL of T cell media containing 200 nM SiR-DNA nucleus stain. SiR-DNA stain does not require washing out.
2. Plate the DN3a and DN3b cells separately in optical bottom petri dishes. Incubate the dishes at 37 °C in a CO_2 incubator for 2 h to allow the cells to adhere to the dish surface.
3. Prepare the LSCM for imaging. Set the excitation wavelength to 488 nm and collect the emission of MitoView Green at 523 nm. Additionally, use a 640 nm excitation wavelength to collect the emission of SiR-DNA at 674 nm (Fig. 2a).
4. Use a 60× oil objective (with a numerical aperture of 1.35) for imaging. Ensure that the cells are properly in focus before proceeding to the next step.
5. Enable the 3D imaging modality of the LSCM. Set the optical distance between image slices to 0.5 μm, capturing a series of images above and below the cell volume. This will provide volumetric information about the cell in three dimensions (*see* **Note 6**).
6. Acquire the image stack using the LSCM. Ensure that the scanning speed, frame averaging, and any other relevant acquisition parameters are appropriately set for optimal image quality.
7. If desired, utilize the software provided with the LSCM to create a 3D render of the cell. Follow the software instructions to generate a visual representation of the cell in three dimensions. Alternatively, export the image stack for further processing and analysis in MATLAB or other image analysis software (Fig. 2b).

3.6 Image Processing and Segmentation

3.6.1 Background Removal and Image Visualization

1. Load the images of both nucleus and mitochondria to be processed into MATLAB using the "imread" function.
2. For a stack of images, use a "for" loop to read all the images.
3. Since each image is in RGB format, use sum("Image," 3) to add all channels and create a 2D image. Do this for both nucleus and mitochondria images.
4. Calculate the threshold value for background noise removal. In this protocol, it is set as 40% of the maximum intensity pixel in the image.
5. Apply the threshold to the green channel using the syntax: Image2D(Image2D < threshold) = 0 (*see* **Note** 7).
6. Store the processed image in a 3D matrix.
7. To create a maximum intensity projection of a 3D stack in MATLAB, use the "max" function with the "[]" and "3" arguments to take the maximum intensity projection along the third dimension (which corresponds to the z-axis).
8. Use the command "imshow()" to display the maximum intensity projection of individual channels. Pseudo colors can be added to the images.
9. To visualize the 3D render of the cell using the volumeViewer app of MATLAB. The colors of the 3D render can be adjusted using the GUI of the volumeViewer app.

The workflow for image loading, denoising, and visualization is outlined in Fig. 3. Detailed MATLAB code for implementing this workflow can be found in the Supplementary materials, available on our GitHub repository.

3.6.2 Cell Segmentation and Automated Image Analysis

1. Combine the 2D maximum projections of the nucleus and mitochondria channels to obtain a single 2D image. Combining both channels will provide a holistic cross-sectional area of the cell.
2. Convert the denoised image into a 2D feature matrix, representing the x- and y-coordinates of each nonzero pixel, intensity of the nucleus, and intensity of the mitochondria at each pixel.
3. Apply the DBSCAN algorithm to identify clusters of pixels that are spatially close and separated from other clusters by low-density regions.
4. Choose the hyperparameters needed for cluster assignment, namely, the search radius (r) and the minimum number of points required to form a cluster (MinPts).
5. To find the optimal values for r and MinPts, use a training image that has been manually labeled with cell boundaries. Use a Monte Carlo approach to explore different combinations

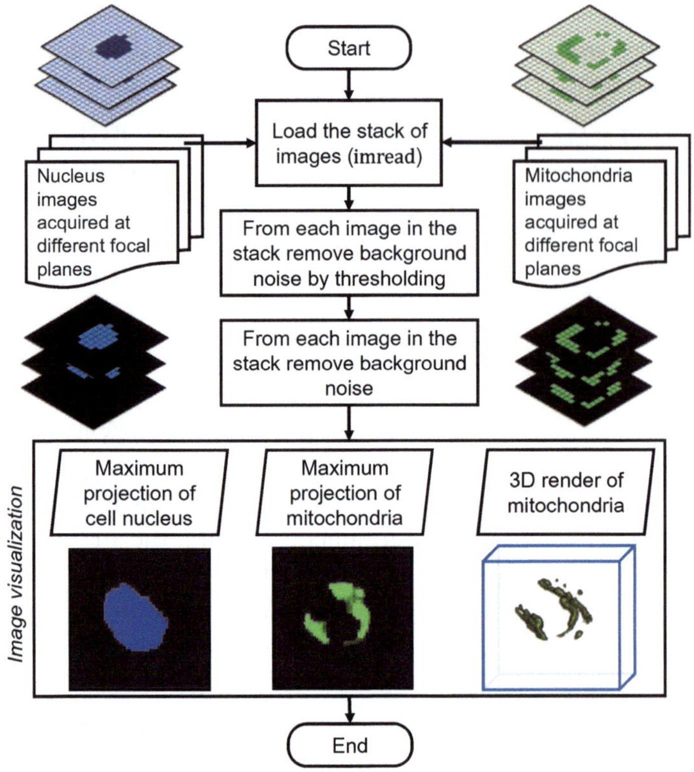

Fig. 3 Workflow for background removal and visualization of 3D images

of r and MinPts and find the values that yield the best segmentation (matching the number of cells in the ground truth and segmented output).

6. In this protocol, $r = 3$ and MinPts = 25 were identified as the optimum parameter values. To explore different parameter values, use generate random parameter combinations. For example, $r = 2 + (6–2)$*rand(1) will generate random search radius between 2 and 6. Similarly, MinPts = randi(25) will generate random integer value of minimum points required to form a cluster between 1 and 25.
7. Once the optimal values of r and MinPts have been determined for a specific type of image, apply the DBSCAN algorithm to detect cells in other images of the same type.
8. Remove the noise pixels that have cluster index (idx) as −1.
9. The algorithm detects cells as clusters, for each cluster, find the centroid coordinates and total mitochondria intensity (stored in the fourth row of the feature matrix). The total mitochondria intensity correlates to mitochondrial content.
10. Plot the segmentation results as a scatter plot of each cluster.

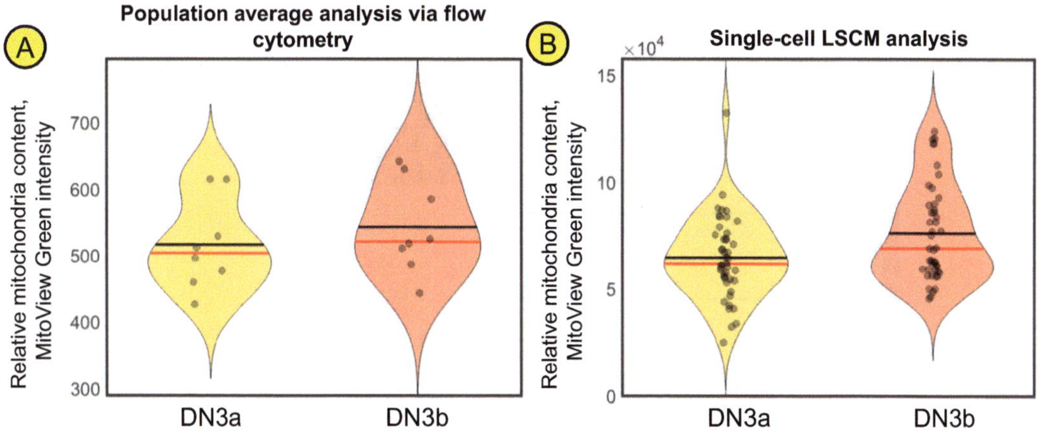

Fig. 4 Comparison of mitochondria content in representative DN3a and DN3b populations from multiple experiments using flow cytometry and LSCM imaging (with automated image analysis and cell segmentation). (**a**) Flow cytometry: Mean values of MitoView Green fluorescence from eight independent experiments are plotted for both DN3a and DN3b cell populations. Each data point represents the mean fluorescence intensity obtained from a single flow cytometry experiment. (**b**) LSCM imaging with automated image analysis: Total MitoView Green intensity in individual DN3a or DN3b cells is depicted by each data point. A total of 45 cells were considered for each population (black and red horizontal lines indicate mean and median values, respectively)

11. Plot a violin plot of the total mitochondrial intensity values from each cell, optionally superimposing the intensity data for better visualization. Compare with the results obtained from flow cytometry analysis (Fig. 4).
12. The volume of mitochondria within an individual cell can be determined by estimating the number of pixels within the cell that exhibit fluorescent intensity above background threshold. By converting the number of pixels to volume units (μm^3), the volume of mitochondria can be quantified (*see* **Note 6**).

The workflow for DBSCAN based image segmentation is presented in Fig. 5. Detailed MATLAB code for implementing this workflow can be found in the Supplementary materials, available on our GitHub repository.

3.7 Live Cell Imaging

1. Prepare the Chamber Slide:
   (a) Seed 2 × 10^4 OP9-DL1 stromal cells in an eight-well chamber slide with PDMS cell paddocks at least 4 h prior to imaging (*see* **Note 8**).
   (b) Place the cells in the middle four wells of the chamber slide.

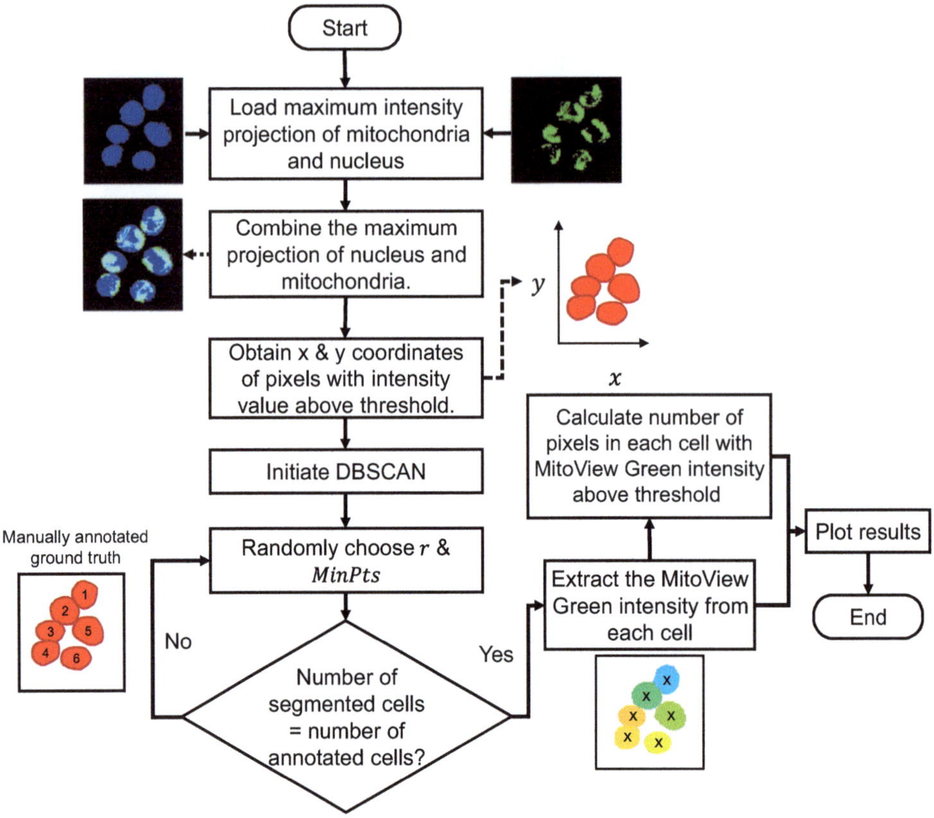

Fig. 5 Workflow for segmentation of cells using DBSCAN

2. Seed T cells:
   (a) Aspirate the OP9-DL1 media from the OP9-DL1 seeded wells of the chamber slide.
   (b) Suspend both DN3a and DN3b populations in T cell media containing 200-nM SiR-DNA nucleus stain.
   (c) Seed 2×10^4 DN3a cells per well in two of the middle wells of the 8 wells in the chamber from the suspension of DN3a cells (Fig. 6a–d).
   (d) Add 2×10^4 DN3b cells per well to the remaining two wells.
3. Place the chamber slide with the seeded cells in a temperature-controlled chamber, mounted on top of a SDCM. Maintain the chamber at 37 °C and 10% CO_2 (Fig. 6e).
4. Excite the sample using 488 nm and 640 nm lasers. Capture the emission of MitoView Green at 523 nm and SiR-DNA at 674 nm, respectively.
5. Use a 20× air objective (with a numerical aperture of 0.75) on the SDCM for imaging.

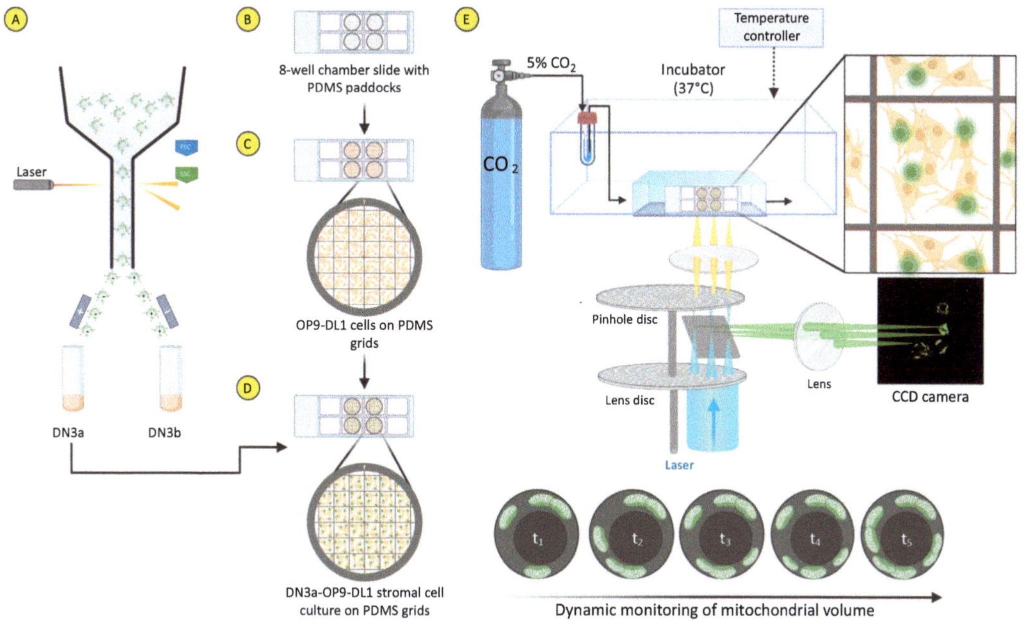

Fig. 6 Workflow for dynamic monitoring of mitochondrial content using SDCM. (**a**) Pure populations of DN3a and DN3b cells are obtained through cell sorting. (**b**) The PDMS paddocks are placed in the middle four wells of an 8-well chamber slide. The PDMS paddocks create a grid-like pattern that restricts the movement of T cells within individual microgrids [8, 23, 24]. (**c**) OP9-DL1 cells are seeded on the PDMS paddocks 4 h before seeding T cells. (**d**) The sorted T cells are seeded on OP9-DL1 cells. (**e**) The setup for live cell imaging in SDCM involves placing the chamber slide on the microscope stage, ensuring that at least one microgrid square is fully within the field of view. This allows for continuous monitoring of mitochondrial content dynamics in the confined T cell population

6. Record multiple stage positions every 3 min for 48 h, ensuring long-term time-lapse imaging of the cells.
7. To perform volumetric imaging, mark the focal plane above and below the cell volume, capturing images at multiple z-planes (*see* **Note 9**).
8. Mark multiple positions to acquire time-lapse from multiple microgrids (*see* **Note 9**).

3.8 Cell Tracking for Dynamic Information

1. Using the protocol mentioned in Subheading 3.6.1, obtain maximum intensity projections of both nucleus and mitochondria at each time point (Fig. 7a, b).
2. At each time point, segment the combined 2D image using DBSCAN algorithm.
3. Save the values of x- and y-coordinates of each cell's centroid along with total mitochondrial intensity as a cell feature matrix. At each time point, save the cell feature matrices as data cells by bracket assignment.

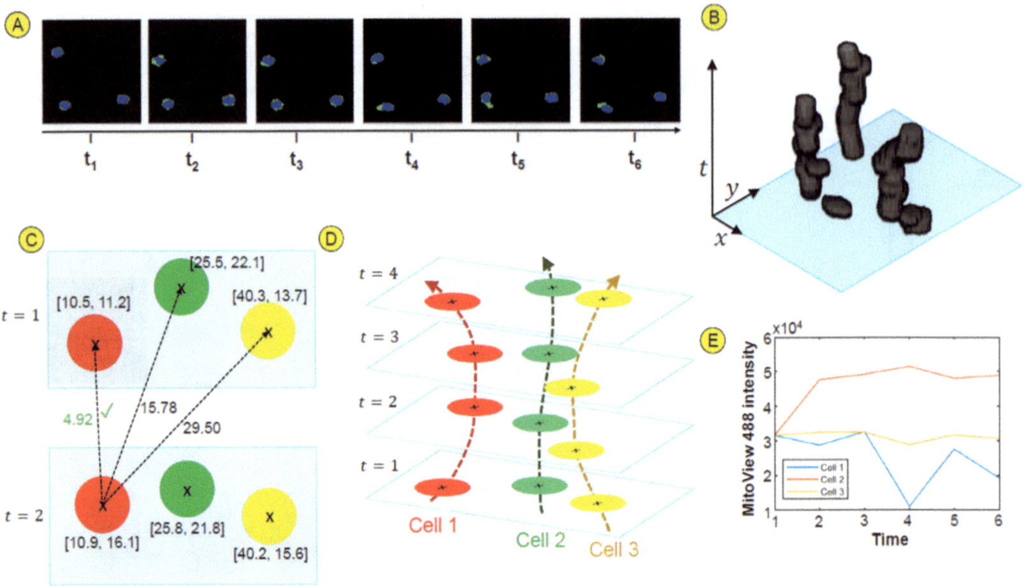

Fig. 7 Cell tracking to obtain dynamics of mitochondrial content: (**a**) Representative time-lapse of three cells at six time points with a 3 min interval, depicting temporal changes in mitochondrial content and cell movement in the x- and y directions. (**b**) 3D visualization of the trajectories of the three cells, providing a spatial representation of their movement patterns. (**c**) To track the motion of a single cell between two-time frames, the Euclidean distance between the cell of interest in the current frame and the cells detected in the previous frame is calculated. The cell in the previous frame with the least Euclidean distance is assigned as the predecessor to the cell of interest in the current frame, establishing the cellular lineage and tracking continuity. (**d**) Trajectory of three representative cells after performing segmentation and cell tracking, demonstrating the path followed by each cell over time. (**e**) Time-lapse of MitoView Green intensity of three represented cells obtained from segmentation and centroid tracking

4. Set the initial values of cell centroid coordinates and mitochondrial intensity using the cell features at the first time point.
5. For time points greater than 1, calculate the Euclidean distance between the position vectors of all cells in the two-time frames. Identify cells with the least distance as the same cells and track their trajectories by linking their position vectors (Figs. 7c, d and 8).
6. Repeat **steps 1–5** for all subsequent time frames using a "for" loop.
7. Obtain the trajectory of each cell by linking its position vectors across all time frames.
8. Visualize the time-lapse of MitoView intensity of the segmented cells (Fig. 7e).

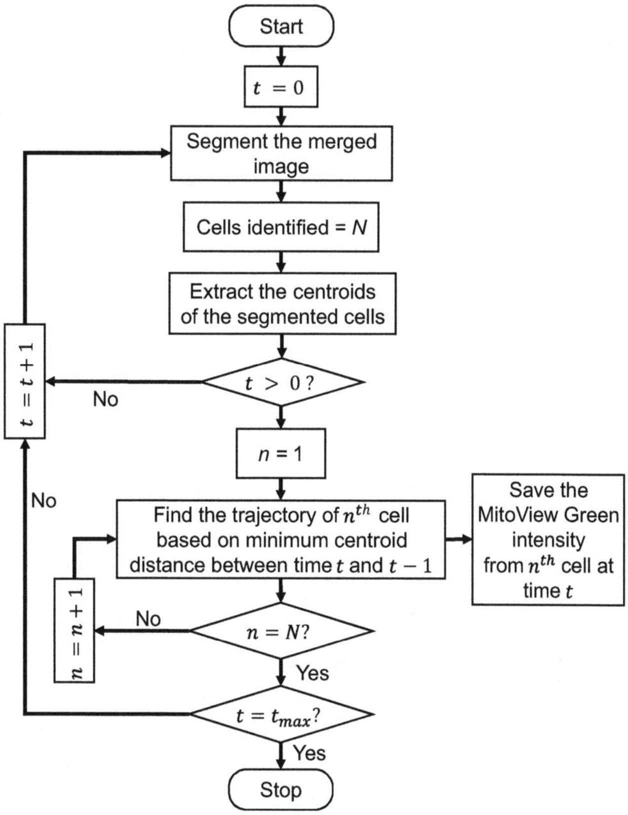

Fig. 8 Workflow for cell tracking to obtain the time-lapse of mitochondrial content in individual cells

4 Notes

1. *Tradeoff between LSCM and SDCM.* LSCM is known for its high-resolution imaging capabilities, particularly in 3D snapshot imaging. It uses a single-point scanning mechanism to excite fluorescence and collect emitted light from the sample. This scanning approach allows for precise control over the excitation and detection, resulting in excellent image quality with high resolution and optical sectioning. LSCM is ideal for capturing detailed structures and fine cellular features in a single time point, making it suitable for high-resolution 3D snapshot imaging. On the other hand, SDCM is preferred for long-term imaging experiments due to its rapid imaging speed and low phototoxicity. SDCM uses a spinning disc confocal unit, which consists of a spinning disk with multiple apertures that generate an array of excitation spots on the sample. These multiple spots enable faster image acquisition compared to LSCM, making it suitable for capturing dynamic processes in real-time or over extended periods. Additionally, the reduced

light exposure per frame in SDCM minimizes phototoxicity, allowing for longer-term imaging without significantly affecting cell viability. However, one trade-off of SDCM is its relatively lower spatial resolution compared to LSCM. The spinning disc approach sacrifices some spatial resolution in exchange for faster imaging speed and reduced phototoxicity. Consequently, while SDCM may not offer the same level of intricate details as LSCM in a single snapshot, it enables capture of dynamic processes while ensuring cell viability in long-term imaging experiments [10].

2. *The choice to use MitoView Green.* In this study, MitoView Green was chosen as the mitochondrial stain due to its membrane potential independence, allowing its use in both live and fixed cells. The green channel was selected for capturing MitoView Green emission, enabling the utilization of the red channel for MitoSox Red™ [19] staining if mitochondrial superoxide imaging was desired. Additionally, the far-red channel was used for SiR-DNA staining. To avoid potential cell damage caused by UV excitation, Hoechst 33342 staining for the nucleus was omitted from the protocol [20]. Alternatively, MitoTracker Green [21], MitoTracker Red [19], and MitoTracker FarRed [22] dyes can be used, but this would necessitate reevaluation of the staining combinations in the remaining emission channels.

3. *Advantages of using DBSCAN.* DBSCAN is a versatile clustering algorithm that is particularly well suited for segmenting images with irregular or complex structures. Unlike certain other methods, such as k-means, which assume circular cluster shapes, or Gaussian mixture models, which assume Gaussian distribution patterns, DBSCAN does not impose any specific restrictions on cluster shapes or distributions. This flexibility allows DBSCAN to effectively identify clusters of arbitrary shapes, making it highly adaptable to diverse image datasets.

   Another advantage of DBSCAN is its relatively low sensitivity to parameter settings compared to methods like k-means or Gaussian mixture models. DBSCAN's parameters, such as the minimum number of points in a cluster and the distance threshold, have intuitive interpretations and can be adjusted to accommodate different datasets without significantly impacting the segmentation results. This adjustability makes DBSCAN a more robust choice for image segmentation, as it can handle varying data characteristics and adapt to different segmentation requirements.

   Furthermore, DBSCAN proves to be a robust and reliable method for image segmentation, particularly in scenarios where annotated images are not readily available to train deep learning networks. Deep learning approaches often require large amounts of accurately labeled training data to achieve satisfactory segmentation performance. However, annotating images can be time-consuming, expensive, or impractical in certain situations.

4. *Limitations of DBSCAN.* While DBSCAN offers several advantages for image segmentation, it also has some limitations. DBSCAN assumes that clusters have similar densities and struggles to handle datasets with clusters of significantly different densities. Another limitation of DBSCAN-based image segmentation is its tendency to over-segment and treat different mitochondria within the same cell as separate clusters. This can result in fragmented and inaccurate segmentation results, as individual mitochondria may be erroneously identified as distinct clusters. To address this limitation, one approach is to consider the contextual information of the cell as a whole. By combining the information of the nucleus and mitochondria, it is possible to create a more comprehensive representation of the cell structure. By segmenting the cell as a whole, it becomes possible to capture the overall structure and organization of the mitochondria within the context of the nucleus. This can lead to more meaningful and biologically relevant segmentation results, enabling better analysis and understanding of cellular processes.

5. *Compensation and control samples.* It is important to prepare single-stain and unstained samples to compensate for spectral overlap. The single-stained samples of CD4, CD8, CD44, CD25, and CD28 antibodies, and MitoView Green, will help determine the extent of spill over into other detection channels. These controls allow the user or software to calculate the compensation matrix, which is a set of correction factors applied to each detection channel to remove the spectral overlap. The unstained control is used to measure the background fluorescence, autofluorescence, and any nonspecific binding of the fluorochromes. It helps establish the baseline signal and provides a reference point for comparison with the stained samples.

6. *Volumetric information from confocal imaging.* To accurately estimate the volumetric content of mitochondria in a cell, a 3D imaging approach can be employed, capturing a series of images along the height of the cell. By analyzing these images, the cross-sectional area of the mitochondria can be determined for each individual stack. The mitochondrial volume can then be calculated by summing the pixel intensities that exhibit fluorescence above a specified threshold. To perform this calculation, it is crucial to consider the known dimensions of each pixel in micrometers (μm) and the predetermined distance between individual focal planes set by the user. By multiplying

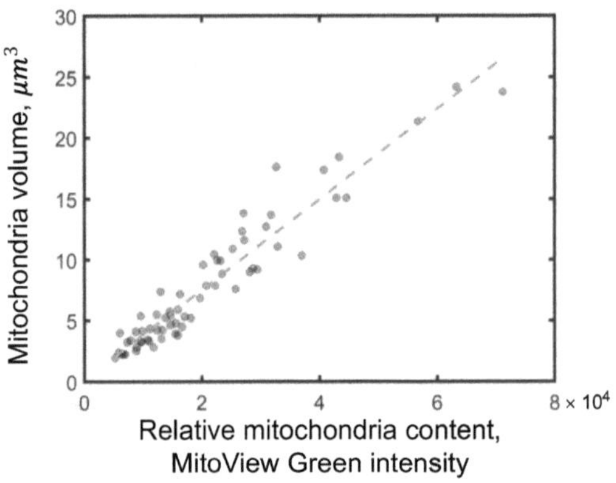

Fig. 9 Scatter plot demonstrating the linear correlation (R = 0.9613) between mitochondrial volume and MitoView Green intensity individual cells (64 cells from three independent experiments)

the sum of the fluorescence-intense pixels by the pixel resolution (μm/pixel) and the height of the z-scan, the volumetric information can be accurately obtained. In our study, we found the calculated mitochondrial volume to be linearly correlated to the MitoView Green florescence (Fig. 9). To visualize the resulting data in a comprehensive manner, 3D rendering software, such as the volumeViewer in MATLAB, can be utilized. These software applications interpolate the surface area between the individual stacks, resulting in a smooth and visually interpretable 3D structure.

However, it is important to consider the trade-off between 3D resolution and photobleaching. While acquiring a 3D stack along the cell height with a large number of cross-sections can provide a high-resolution 3D render, it is necessary to be cautious of excessive laser excitation. Excessive laser exposure may lead to photobleaching of the stain, potentially affecting the accuracy and reliability of the measurements.

7. *Background removal.* The threshold value can be adjusted to a different percentage of the maximum intensity pixel, or alternatively, the values of background intensity can be manually inputted into the algorithm. MATLAB's Image Processing Toolbox offers several noise removal options, including the following:
   - Gaussian filtering: imgFiltered = imgaussfilt(image, sigma).
   - Median filtering: imgFiltered = medfilt2(image, [m n]).
   - Wiener filtering: imgFiltered = wiener2(image, [m n]).
   - Wavelet denoising: imgFiltered = wdenoise2(image, threshold, waveletName).
   - Total Variation Denoising: imgFiltered = denoiseTV (image, lambda, nIter).

The parameters of these methods can be adjusted to achieve the best background removal results. In this protocol, threshold-based noise removal is demonstrated due to its ease of application and simple tuning process.

8. *Rationale of using PDMS paddocks.* T cells are highly mobile in nature, which can pose challenges in tracking their movement during imaging experiments. To address this, the use of PDMS paddocks can be beneficial. These paddocks create a grid-like pattern that confines the movement of T cells within individual square-shaped microgrids [8, 14, 23, 24]. This allows for the tracking of T cells within that specific grid throughout the duration of the experiment, facilitating the study of cell migration dynamics within defined spatial boundaries. Before seeding the cells on the PDMS grids, they are placed in the chambers or petri dish and then rendered hydrophilic using air plasma or UV treatment. When setting up the sample in SDCM, it is important to ensure that at least one microgrid square is fully within the field of view. Since the field of view can be microscope specific, user may choose the objective-digital zoom configuration that works best for their experimental system.

9. *Fine tuning z-stack height and number of positions in live cell imaging.* During volumetric time-lapse imaging experiments, it is critical to account for the gradual movement of the cell throughout the duration of the study. Microscope focal plane drift and cell displacement along the z-direction can have a substantial impact on the accuracy and reliability of the acquired data. Moreover, the use of autofocus functionality may be impractical when imaging multiple positions simultaneously. To overcome these challenges, it is recommended to implement specific strategies.

   One approach is to provide an additional margin in the z-direction or designate z-positions for z-stack imaging that extend beyond the actual height of the cell. This allows for tracking small cell movements over extended periods. By capturing a broader z-range, any cell displacements within that range can still be adequately documented and analyzed. However, it is important to exercise caution when increasing the z-stack height, as a larger stack height may also result in increased laser exposure, which can have implications for cell viability.

   Another crucial consideration is the potential for erroneous stage shift when imaging multiple positions. While modern microscopes typically possess robust motorized stage control, the likelihood of erroneous stage drift or shift rises with the number of positions and distance between the positions being imaged. Therefore, researchers must carefully optimize the number of xy-positions along with the desired stack height that can be effectively imaged over the long term. Striking the right balance between imaging multiple positions and minimizing stage-related issues is vital for obtaining accurate and reliable volumetric time-lapse data. In the described protocol, the use of the middle four wells of the 8-well chamber slide is a strategic approach to address these concerns. By imaging only the middle four wells, the positions being imaged are not too far apart, reducing the chances of significant stage drift between positions. Additionally, this approach ensures that the microscope objective does not hit the stage or potentially disrupt the sample, which could occur when imaging positions close to the edges of the slide.

Acknowledgments

This work was done on Wurundjeri land of the Kulin nation, and we pay our respects to the Elders past and present. Authors would like to acknowledge BioRender for their graphic design software.

Supporting Information The supporting information, including additional data, code, and figures, is available on GitHub at https://github.com/VaibhavDhyani/Sample_codes_for_image_segmentation. Please refer to this link for further details.

Funding

National Health and Medical Research Council grant APP1099140 (SMR).
Flow cytometry was supported by a grant for the FACS Aria from the L.E.W Carty Charitable Fund.
CASS Foundation (CASS medicine/science grant number: 7818, MC).

References

1. Afroz SF, Condon ND, Sweet MJ, Kapetanovic R (2022) Quantifying regulated mitochondrial fission in macrophages. Methods Mol Biol 2523:281–301. https://doi.org/10.1007/978-1-0716-2449-4_18
2. Piconese S, Campello S, Natalini A (2020) Recirculation and residency of T cells and Tregs: lessons learnt in Anacapri. Front Immunol 11. https://doi.org/10.3389/fimmu.2020.00682
3. Shyer JA, Flavell RA, Bailis W (2020) Metabolic signaling in T cells. Cell Res 30(8): 649–659. https://doi.org/10.1038/s41422-020-0379-5
4. Sun V, Sharpley M, Kaczor-Urbanowicz KE et al (2021) The metabolic landscape of Thymic T cell development in vivo and in vitro. Front Immunol 12. https://doi.org/10.3389/fimmu.2021.716661

5. Zhang M, Lin X, Yang Z et al (2022) Metabolic regulation of T cell development. Front Immunol 13. https://doi.org/10.3389/fimmu.2022.946119
6. Corrado M, Samardžić D, Giacomello M et al (2021) Deletion of the mitochondria-shaping protein Opa1 during early thymocyte maturation impacts mature memory T cell metabolism. Cell Death Differ 28(7):2194–2206. https://doi.org/10.1038/s41418-021-00747-6
7. Chann AS, Russell SM (2019) An integrated transcriptional switch at the β-selection checkpoint determines T cell survival, development and leukaemogenesis. Biochem Soc Trans 47(4):1077–1089. https://doi.org/10.1042/bst20180414
8. Charnley M, Russell SM (2017) Imaging asymmetric T cell division. In: Baldari CT, Dustin ML (eds) The immune synapse: methods and protocols. Springer, New York, pp 383–398. https://doi.org/10.1007/978-1-4939-6881-7_23
9. Cox G (2002) Biological confocal microscopy. Mater Today 5(3):34–41. https://doi.org/10.1016/S1369-7021(02)05329-4
10. Elliott AD (2020) Confocal microscopy: principles and modern practices. Curr Protoc Cytom 92(1):e68. https://doi.org/10.1002/cpcy.68
11. Jonkman J, Brown CM (2015) Any way you slice it-a comparison of confocal microscopy techniques. J Biomol Tech 26(2):54–65. https://doi.org/10.7171/jbt.15-2602-003
12. DeCamp SJ, Tsuda VMK, Ferruzzi J et al (2020) Epithelial layer unjamming shifts energy metabolism toward glycolysis. Sci Rep 10(1):18302. https://doi.org/10.1038/s41598-020-74992-z
13. Polo M, Alegre F, Moragrega AB et al (2017) Lon protease: a novel mitochondrial matrix protein in the interconnection between drug-induced mitochondrial dysfunction and endoplasmic reticulum stress. Br J Pharmacol 174(23):4409–4429. https://doi.org/10.1111/bph.14045
14. Charnley M, Allam AH, Newton LM et al (2023) E-cadherin in developing murine T cells controls spindle alignment and progression through b-selection. Sci Adv 9(3): eade5348. https://doi.org/10.1126/sciadv.ade5348
15. Bishop EL, Gudgeon NH, Mackie GM et al (2022) 1,25-Dihydroxyvitamin D3 suppresses CD4+ T-cell effector functionality by inhibition of glycolysis. Immunology 166(3): 299–309. https://doi.org/10.1111/imm.13472
16. Vaeth M, Maus M, Klein-Hessling S et al (2017) Store-operated Ca(2+) entry controls clonal expansion of T cells through metabolic reprogramming. Immunity 47(4):664–679.e666. https://doi.org/10.1016/j.immuni.2017.09.003
17. Bandyopadhyay SK, Paul TU (2013) Segmentation of brain tumour from MRI image analysis of k-means and dbscan clustering. Int J Res Eng Sci 1(1):48–57
18. mehdi Cherrat E, Alaoui R, Bouzahir H (2019) Improving of fingerprint segmentation images based on K-means and DBSCAN clustering. Int J Electr Comput Eng (IJECE) 9(4): 2425–2432
19. Akkaya B, Roesler AS, Miozzo P et al (2018) Increased mitochondrial biogenesis and reactive oxygen species production accompany prolonged CD4(+) T cell activation. J Immunol 201(11):3294–3306. https://doi.org/10.4049/jimmunol.1800753
20. Singh S, Dwarakanath BS, Mathew TL (2004) DNA ligand Hoechst-33342 enhances UV induced cytotoxicity in human glioma cell lines. J Photochem Photobiol B 77(1–3): 45–54. https://doi.org/10.1016/j.jphotobiol.2004.08.009
21. Konjar Š, Frising UC, Ferreira C et al (2018) Mitochondria maintain controlled activation state of epithelial-resident T lymphocytes. Sci Immunol 3(24). https://doi.org/10.1126/sciimmunol.aan2543
22. Scharping NE, Menk AV, Moreci RS et al (2016) The tumor microenvironment represses T cell mitochondrial biogenesis to drive Intratumoral T cell metabolic insufficiency and dysfunction. Immunity 45(2): 374–388. https://doi.org/10.1016/j.immuni.2016.07.009
23. Charnley M, Ludford-Menting M, Pham K, Russell SM (2019) A new role for Notch in the control of polarity and asymmetric cell division of developing T cells. J Cell Sci 133(5). https://doi.org/10.1242/jcs.235358
24. Day D, Pham K, Ludford-Menting MJ et al (2009) A method for prolonged imaging of motile lymphocytes. Immunol Cell Biol 87(2):154–158. https://doi.org/10.1038/icb.2008.79

Chapter 13

Reconstructing Signaling Networks Using Biosensor Barcoding

Suyang Wang, Wei-Yu Chi, Gabriel Au, Cheng-Chieh Huang, Jr-Ming Yang, and Chuan-Hsiang Huang

Abstract

Understanding how signaling networks are regulated offers valuable insights into how cells and organisms react to internal and external stimuli and is crucial for developing novel strategies to treat diseases. To achieve this, it is necessary to delineate the intricate interactions between the nodes in the network, which can be accomplished by measuring the activities of individual nodes under perturbation conditions. To facilitate this, we have recently developed a biosensor barcoding technique that enables massively multiplexed tracking of numerous signaling activities in live cells using genetically encoded fluorescent biosensors. In this chapter, we detail how we employed this method to reconstruct the EGFR signaling network by systematically monitoring the activities of individual nodes under perturbations.

Key words Genetically encoded fluorescent biosensors, Live cell imaging, Signaling network, Deep learning, Receptor tyrosine kinases (RTKs), Epidermal growth factor receptor (EGFR), Multiplexing, Small-molecule inhibitors

1 Introduction

Numerous biochemical activities are responsible for coordinating cellular functions in both space and time. To track these activities in live cells, many genetically encoded fluorescent biosensors have been developed utilizing fluorescent proteins (FPs). These biosensors allow for the monitoring of various cellular activities, such as signaling and metabolic processes [1]. However, the limited spectral space and broad emission spectra of FPs have prevented the imaging of more than a few biosensors in a single experiment, despite attempts to increase multiplexity. One such attempt includes expanding the color spectrum of biosensors or targeting them to different intracellular sites [2]. However, these strategies

Authors Suyang Wang and Wei-Yu Chi have equally contributed to this chapter.

Christoph Wülfing and Robert F. Murphy (eds.), *Imaging Cell Signaling*, Methods in Molecular Biology, vol. 2800, https://doi.org/10.1007/978-1-0716-3834-7_13,

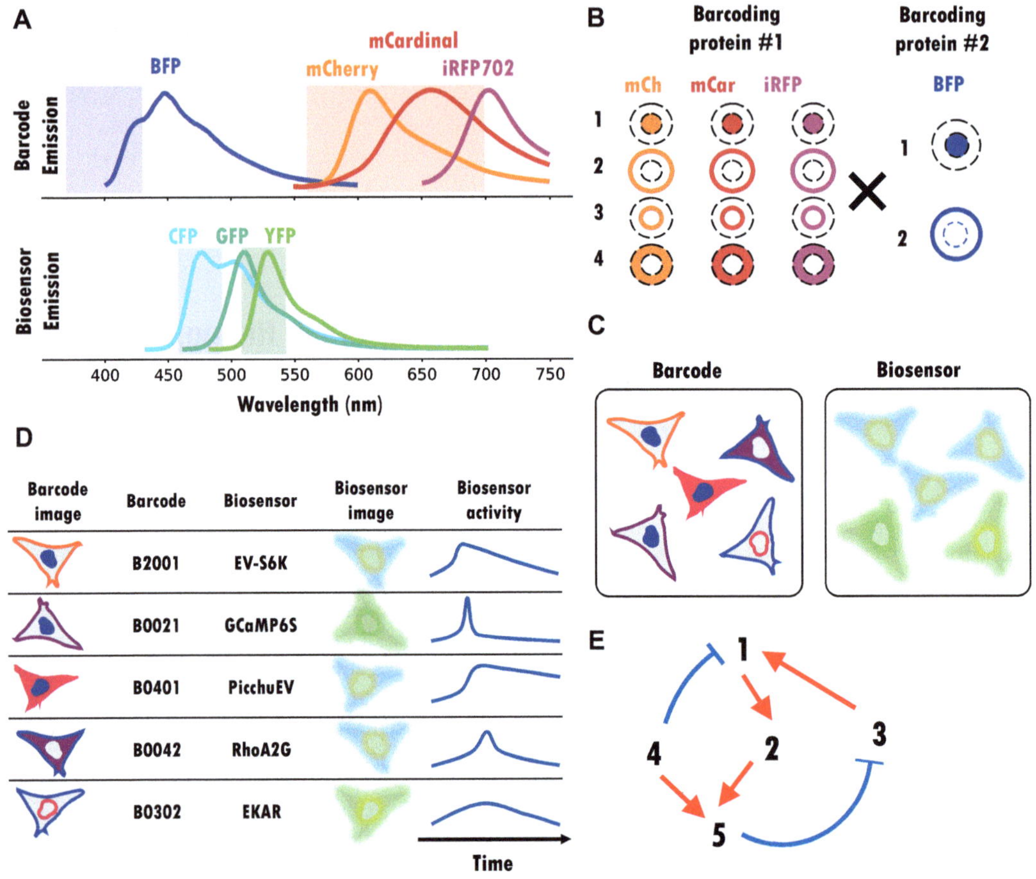

Fig. 1 Overview of network reconstruction using biosensor barcoding (**a**) Emission spectrums of barcoding proteins and biosensors. Shaded regions indicate the range of collection. (**b**) Generation of barcodes from combinations of two barcoding proteins: the first based on a red FP and the second BFP. (**c**) Cells are mixed for simultaneous imaging of barcodes and biosensors. (**d**) The responses to biosensors to inhibitors are obtained from time-lapse images of biosensors, the identity of which can be determined by the barcodes expressed in the same cells. (**e**) The interactions between the nodes can be deduced from the responses to inhibitors

have limited success in expanding the number of activities to be monitored, as no more than five to six different biosensors can be imaged at the same time.

To overcome the limitation of imaging only a few biosensors in a single experiment, we recently introduced a method for massively multiplexed biosensor imaging [3, 4]. Our approach, known as "biosensor barcoding," involves labeling cells expressing different biosensors with barcodes consisting of a pair of barcoding proteins—blue or red fluorescent proteins that are targeted to distinct subcellular locations (Fig. 1a, b). Unlike the commonly used cyan, green, or yellow FPs in biosensors, the emission spectra of the barcoding proteins are easily separable. Following cell mixing, spectral images of the barcodes are obtained at the beginning of the experiment, followed by time-lapse imaging of biosensors (Fig. 1c) using a fluorescent microscope equipped with a spectral detector. The barcode present in each cell allows for the identification of the biosensor expressed, which can be achieved using machine learning models. Subsequently, the signals of cells expressing the same biosensor can be combined for further analysis (Fig. 1d). Information from the responses to perturbations can reveal the regulatory structure of signaling networks (Fig. 1e).

Besides improving the efficiency of data acquisition, the biosensor barcoding method has other benefits. Firstly, cells expressing various biosensors exhibit synchronized activities, which simplifies the comparison of different biosensors. Secondly, the capability to label distinct cell populations in the same mixture enables differentiation between cell-autonomous and nonautonomous effects [3].

By enabling massively parallel imaging of biosensors, our method provides a powerful tool for reconstructing signaling networks. Specifically, we can use the method to track the activities of individual nodes in the signaling network and observe their responses when the nodes are individually blocked using small molecule inhibitors. In this chapter, we demonstrate this strategy using the EGFR signaling network as an example (Fig. 2), but the strategy can be easily applied to the study of other signaling networks.

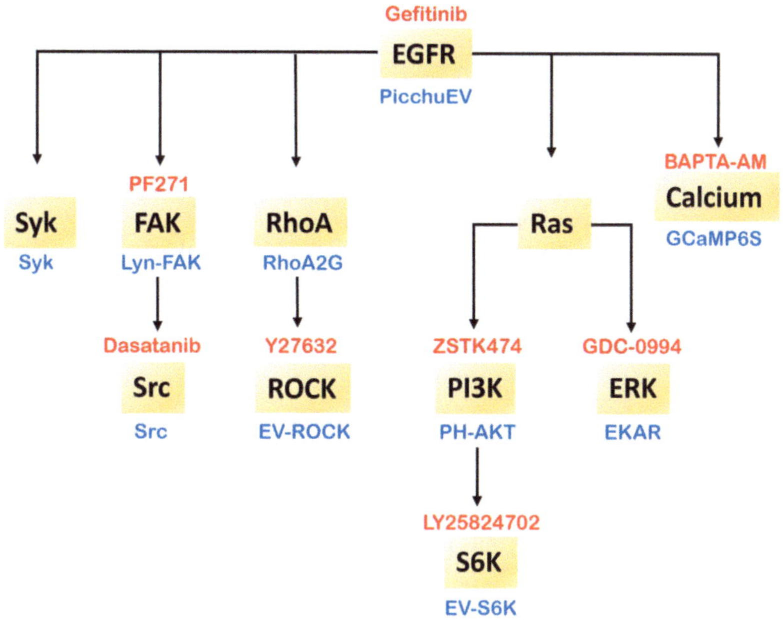

Fig. 2 Biosensors and inhibitors for the EGFR-signaling network. The biosensors used in this chapter are shown in blue and inhibitors in red

2 Materials

2.1 Cell Culture and Transfection

1. HeLa cells are purchased from ATCC (CCL2).
2. Culture medium: DMEM, with L-glutamine and high glucose, supplemented with 10% FBS, 1-mM sodium pyruvate and 1X MEM nonessential amino acids.
3. Imaging medium: DMEM, with L-glutamine and high glucose, HEPES, no phenol red, supplemented with 10% FBS, 1-mM sodium pyruvate and 1X MEM nonessential amino acids.
4. GenJet™ In Vitro DNA Transfection Reagent (Ver. II).

2.2 Plasmids

Barcoding proteins used in this chapter include the following (*see* Table 1 for an explanation of the barcoding protein nomenclature):

1. D1: BFP-NLS, Addgene #184458
2. D2: BFP-CAAX, Addgene #184459
3. B1: mCardinal-NLS, Addgene #184450
4. B3: mCardinal-LaminB1, Addgene #184452

Table 1
Biosensors and barcodes

Biosensor name	Target	Barcoding proteins	Barcode
PicchuEV	EGFR	B4D1	0401
Src biosensor	Src	E4D1	4001
Lyn-FAK	FAK	E3D1	3001
EV-ROCK	ROCK	C1D2	0012
GCaMP6S	Calcium	C2D1	0021
PH-AKT	PI3K	B1D1	0101
EKAR	ERK	B3D2	0302
EV-S6K	S6K	E2D1	2001
Syk biosensor	Syk	E2D2	2002
RhoA2G	RhoA	C4D2	0042

The biosensor and a pair of barcoding proteins are transfected into cells separately. Each barcode consists of a BFP (code: D) targeted to either the nuclear or plasma membrane, and one of the three red FPs: mCherry (E), mCardinal (B),or iRFP702 (C) targeted to one of four subcellular locations: the nucleus (1), plasma membrane (2), nuclear membrane (3), or cytosol (4). The barcode digits, from left to right, indicate the subcellular location of the targeted FPs: mCherry, mCardinal, iRFP702, and BFP, with 0 denoting not expressed. For instance, barcoding protein combination E2D1 represents mCherry at location 2 (plasma membrane) and BFP at location 1 (nucleus), with the corresponding barcode 3001

5. B4: mCardinal-NES, Addgene #184453
6. C1: iRFP702-NLS, Addgene #184454
7. C2: iRFP702-CAAX, Addgene #184455
8. C4: iRFP702-NES, Addgene #184457
9. E2: mCherry-CAAX, Addgene #186347
10. E3: mCherry-LaminB1, Addgene #186348
11. E4: mCherry-NES, Addgene #186349

Biosensors:

1. PicchuEV [5]
2. RhoA2G, Addgene #40176 [6]
3. Lyn-FAK biosensor, Addgene #78299 [7]
4. Cytosolic Syk Biosensor, Addgene #125729 [8]
5. PH-AKT-GFP, Addgene #51465 [9]
6. Cytoplasmic EKAR (Cerulean-Venus), Addgene #18679 [10]
7. GCaMP6s, Addgene #40753 [11]
8. EV-ROCK [12]
9. Src biosensor, Addgene #78302 [13]
10. EV-S6K [5]

2.3 Chemicals

1. Gefitinib: EGFR inhibitor
2. PF562271: FAK inhibitor
3. ZSTK474: PI3K inhibitor
4. GDC-0994: ERK inhibitor
5. BAPTA: Calcium chelator
6. Y27632: ROCK inhibitor
7. Dasatinib: Src inhibitor
8. LY2584702: S6K inhibitor

See also Table 2.

2.4 Imaging Equipment

1. Zeiss AxioObserver with 780-Quasar confocal module, which is equipped with a 34-channel high-sensitivity gallium-arsenide phosphide (GaAsP) spectral detector controlled by the Zen software.
2. Three laser excitation wavelengths are used: 405 nm for BFP, 458 nm for CFP/GFP/YFP, and 633 nm for the red FPs. The emission signals of BFP, CFP, YFP/GFP, and red FPs are collected in the ranges of 371–430 nm, 458–499 nm, 508–543 nm, and 561–695 nm, respectively (*see* **Note 1**).

Table 2
Inhibitors

Drug name	Target	Stock concentration (mM)	Working concentration (μM)	Source	Identifier
Gefitinib	EGFR	1	1	Cayman	13166
PF562271	FAK	10	1	AdipoGen	SYN-1064
ZSTK474	PI3K	10	1	Cell Signaling	13213
GDC-0994	ERK	10	1	APExBIO Technology	B5817
BAPTA-AM	Calcium	10	10	Selleck	S7534
Y27632	ROCK	10	10	Enzo Life Sciences	ALX-270-333
Dasatinib	Src	10	1	Cayman	11498
LY2584702	S6K	1	1	Selleck	S7698

3. An environmental control chamber maintains the cells at 37 °C during time-lapse imaging. The microscope is also equipped with a motorized stage for multi-position image acquisition, and a Definite Focus module for auto-focusing.

2.5 Image Analysis Software

1. ImageJ/Fiji (https://imagej.net/software/fiji/) [14, 15]
2. Python 3 https://www.python.org
3. Keras https://keras.io/ [16]
4. TensorFlow https://www.tensorflow.org/ [17]
5. Scripts and deep learning models for analysis can be downloaded from: https://github.com/BearHuangLab/Biosensor-barcoding

3 Methods

3.1 Transfection of HeLa Cells with Biosensors and Barcodes

1. Day 1, seed 2×10^5 HeLa cells to ten wells in a 12-well tissue culture plate along with 2 mL of DMEM complete medium. Incubate the cells overnight at 37 °C, 5% CO_2 (*see* **Note 2**).
2. Day 2, replace the medium in each well with 0.75 mL of fresh DMEM complete medium 30 min prior to transfection.
3. For each well, prepare one tube of diluted DNA solution and one with the GenJet™ transfection reagent. First, dilute 0.75 μg of DNA in 38 μL of serum-free DMEM. In each transfection, the DNA mixture contains two barcoding proteins and one biosensor (Table 1). Start with a mass ratio of 1:

1:1 and adjust the ratio if necessary to achieve optimal fluorescence signal (*see* **Note 3**). Next for the transfection reagent, add 2.25 μL GenJet™ Transfection Reagent in 38 μL of serum-free DMEM. Mix by vortexing and spin down briefly. Immediately add the diluted transfection reagent to the diluted DNA solution, and pipette up and down a few times to mix. Leave the mixture at room temperature for 15 min.

4. Add the DNA-GenJet™ complex dropwise to cells in the 12-well plate. Incubate cells at 37 °C, 5% CO_2 overnight.
5. Day 3, collect the cells by removing the medium containing DNA-GenJet™ complex. Rinse each well with DPBS, aspirate, and detach the cells with Accutase. Mix cells from all ten wells together and resuspend with Imaging Medium. Seed 5×10^5 cells in 35-mm glass-bottom dishes, and incubate overnight at 37 °C, 5% CO_2.

3.2 Imaging Barcodes with Zeiss LSM780-FCS Laser Scanning Confocal Microscope

1. Before imaging, replace the medium with fresh Imaging Medium for each dish.
2. Turn on the microscope, halogen lamp, and argon laser. Switch the objective to the 40× oil lens and add a small drop of immersion oil. Place the glass-bottom dish on the stage. Remove the culture dish lid for accessible drug administration during time-lapse imaging experiments. Secure the dish with stage clips.
3. Open the Zen software on the computer and turn on the stage incubator to 37 °C. Select the "Locate" setting on the software and switch on "Transmitted light." Locate cells by adjusting microscope focus.
4. Switch to "Acquisition" and select the "Lambda Mode" tab under "Light Path." For the far red barcodes, select the 633-nm laser as the excitation source and set the pinhole size to 30.2. Set range to 561–695 nm with resolution of 8.9 nm (*see* **Note 4**).
5. To capture images, choose 30 positions containing at least six to ten cells with strong fluorescent signals in each view field. Save the barcode images as *[Experiment ID]*_633Ex.lsm (*see* **Note 5**).
6. To differentiate the red barcodes, perform Linear Unmixing using the reference spectra to generate three separate images (mCherry, mCardinal and iRFP702) for each position (the reference spectra can be obtained using cells expressing individual FPs; see [4] for details). Save the file as *[Experiment ID]* _633Ex_Linear unmixing.lsm.

7. For the BFP barcode, acquire images through "Channel Mode." Select the 405 nm laser as the excitation source and set range to 371–430 nm with pinhole size of 30.2. Save the file as *[Experiment ID]* 405Ex.lsm.

3.3 Imaging Cells Treated with Inhibitors

1. After the barcode images are captured, start time-lapse imaging for biosensor responses. For drug inhibition studies, time-lapse images are taken for each position every 3 min for ten frames in total.
2. On the Zen software, click the "Channel Mode" tab under "Light Path" selections. To capture FRET and KTR biosensors, select the spectrum range of 458–499 nm (for CFP) and 508–543 nm (for YFP and GFP). For the excitation source, use the 458-nm laser and set the pinhole size to 300.6.
3. Select "Time Series." Set the frame rate at 3 min per frame for ten frames, and choose "Definite Focus" as the focusing strategy.
4. Immediately after acquiring three images (frame 3) for all positions, add the inhibitor (*see* Table 2 for the list of inhibitors) by carefully lifting the incubator lid and gently add the reagents to the dish.
5. When imaging is done, save the file as *[Experiment ID]*_FRET.lsm.
6. Save the four files (*see* Table 3) generated from each imaging experiment to a separate folder without other files. This is needed for the macros to correctly identify the files.

3.4 Identifying Barcodes Using ImageJ

1. Download and install necessary programs and files: ImageJ Fiji from https://imagej.net/software/fiji/, Python 3 from https://www.python.org/, and necessary scripts and files developed by our lab from https://github.com/BearHuangLab/Biosensor-barcoding. (An example dataset is

Table 3
Imaging file templates and their purposes

Filename	Purpose
*[Experiment ID]*_405Ex.lsm	For generating barcodes: BFP channel (1-channel)
*[Experiment ID]*_633Ex.lsm	For spectral analysis: 15 bins
*[Experiment ID]*_633Ex_Linear unmixing.lsm	For generating barcodes: RFP channels (3-channel)
*[Experiment ID]*_FRET.lsm	For selecting ROIs and measuring CFP/YFP intensity (2-channel, time-lapse)

contained in the folder "4. Toy Example"; the folder contains a readme.txt which describes how to run the example.)

2. Open Fiji. Open the image file ending with "FRET.lsm" by dragging it onto the ImageJ toolbar.
3. Next, open the ImageJ macro titled "2. Barcode identification/Image Analysis.ijm" for barcode selection.
4. In the macro window, select "Run" and enter the number of positions in the images and time points per position. Click "OK" to continue.
5. After clicking "OK," the "FRET.lsm" file will be split into two image windows: "combined-YFP" and "combined-CFP," and a dialog box will prompt the user to select ROIs. DO NOT click OK until all ROIs are selected.
6. Use the selection tools to circle cells with visible fluorescence in the "combine-YFP" window. Adjust the contrast to see cells of low intensity. Individually outline each cell by using the oval select tool and pressing "T" to record the position of the cell into the ROI manager. Each circled area should include only one cell. Scroll across all time points to make sure that cells remain in the circled region (*see* **Note 6**). Click "OK" on the dialogue box when finished.
7. The macro will then prompt the user to enter the experiment name. Enter a name (e.g., "YYYYMMDD HeLa Cells") and press enter. The macro will proceed to collect data for each cell and write it to files in the experiment folder.
8. Before analysis with the deep learning models, create a list of barcodes to classify by editing the barcode list file (*Barcode identification>Barcode reading GUI>Barcode list.txt*).
9. To automatically classify each fluorescent barcode, run the script "Barcode.py" to open the Barcode Prediction GUI. To run the script, open a command line, and type "python" followed by the path to your script, that is,

   > python ./*path_to_the_folder*/Barcode reading GUI/Barcode.py

   Note that the script requires the python package "tensorflow." To install it (if it is not already installed), the latest version of pip must be installed first, and then pip can be used to install tensor flow. Use the following commands:

   >pip install --upgrade pip

   >pip install tensorflow
10. In the GUI, select and load or confirm the prediction models, the barcode list, barcode image folder ("Barcodes" in experiment folder), and spectral data (***_spectrumdata.txt in experiment folder) (*see* **Note 7**). Edit the output directory if

desired. Click "Predict barcode" to generate a file called "modelpred.csv" containing the predicted barcode for each cell.

11. For best results, single-fluorophore biosensors GCaMP6S and PH-AKT should have their ROIs shrunk to include only the cell after classification. For PH-AKT, only the cytosol should be included. This can be done by opening combined-YFP.tif and ***_FRET_ROISet.zip from the experiment folder in ImageJ, shrinking the ROI of each cell with a single-fluorophore, and recording the intensity *More>Multi-Measure>Ok* in the ROI Manager. The resulting intensity data should be pasted to overwrite the YFP data in the *** _ROIData.csv file (also in experiment folder). This step can also be done using the macro *Barcode identification>"PH-AKT Reprocessing.ijm."*

3.5 Analysis of Biosensor Response

1. Open the Excel template *Analysis>"10Mix_Template_STAR_Protocols.xlsx."*
2. Switch to the sheet "Barcodes" and edit the barcode-biosensor combinations if needed.
3. Switch to the sheet "All_Cells" and copy over the biosensor classification and signal data. From the barcode prediction output (modelpred.csv), copy the columns [Index] and [Thresholded] to the template columns [Position] and [Barcode], respectively. From the file ***_ROIdata.csv in the experiment folder, copy the raw fluorophore data and transpose paste (in Excel *Right Click>Paste Options>Transpose (T)*) into the template columns [YFP1]...[CFP10]. Do not copy any row or column headers from data files. The activities will be calculated automatically.
4. The final result will be in the sheet "Analysis." The activities are normalized to the average of those in the prestimulus levels, and biosensors with "inverted" design (i.e., less FRET or fluorescence when active) have the raw signal inverted so that the displayed signal directly corresponds to activity. In other words, the activities represent YFP/CFP for PicchuEV, RhoA-2G, EV-ROCK, EKAR, and EV-S6K; CFP/YFP for Syk, Src, and Lyn-FAK; cytosolic fluorescence (GCaMP6S); and 1/cytosolic fluorescence (PH-AKT).

3.6 Generation of Interaction Map

By analyzing the biosensor responses to inhibitors, we can deduce how the nodes in the network interact with each other. For instance, when compared to the responses to the vehicle control (DMSO), the use of the FAK inhibitor led to a significant decrease in the activity of FAK, Src, and ERK while increasing the activity of RhoA (Fig. 3a). This suggests that FAK has a positive interaction with Src and ERK but a negative interaction with RhoA (Fig. 3b). To construct a comprehensive interaction map for the EGFR

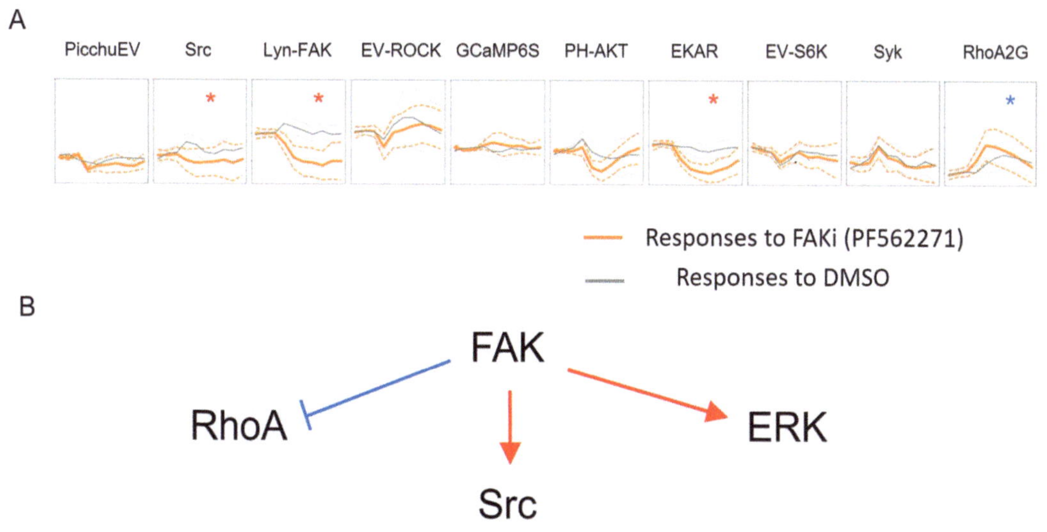

Fig. 3 Inferring interactions from responses to inhibitors (**a**) Responses of biosensors to the FAK inhibitor PF-562271. Red and blue stars indicate statistically significant decrease and increase, respectively, in the activity induced by the FAKi (orange) when compared to the DMSO control (gray). (**b**) Positive (red) and negative (blue) interactions between FAK and other nodes in the network can be inferred from a decreased or increased response, respectively, to the FAK inhibitor

signaling network, we compare the biosensor responses to all inhibitors with those of the vehicle control (Fig. 4a). This analysis comprised the following steps:

1. Calculate the **Total Activity Score (TAS)** of each cell by summing up the activity (from Subheading 3.3, **step 2**) across all time points.
2. For each inhibitor experiment, compare the TAS of cells expressing specific biosensor (inferred from the barcode) with the TAS of the same biosensor in the DMSO control experiment using student's t-test to get the significance (p values).
3. The effects of inhibitors on biosensors can be represented by the matrix of inhibition, in which the value of each element is calculated by: -log(p-value) x direction of change (i.e., +1 for positive effects; −1 for negative effects (Fig. 4b).
4. Generate the interaction diagram based on the matrix of inhibition. As explained above (Fig. 3), positive and negative interactions are assigned for inhibitory and activating effects, respectively (Fig. 4c).

3.7 Concluding Remarks

The interaction map that has been created serves as a valuable tool for future investigations of the EGFR signaling network and for identifying novel interactions within signaling networks of specific

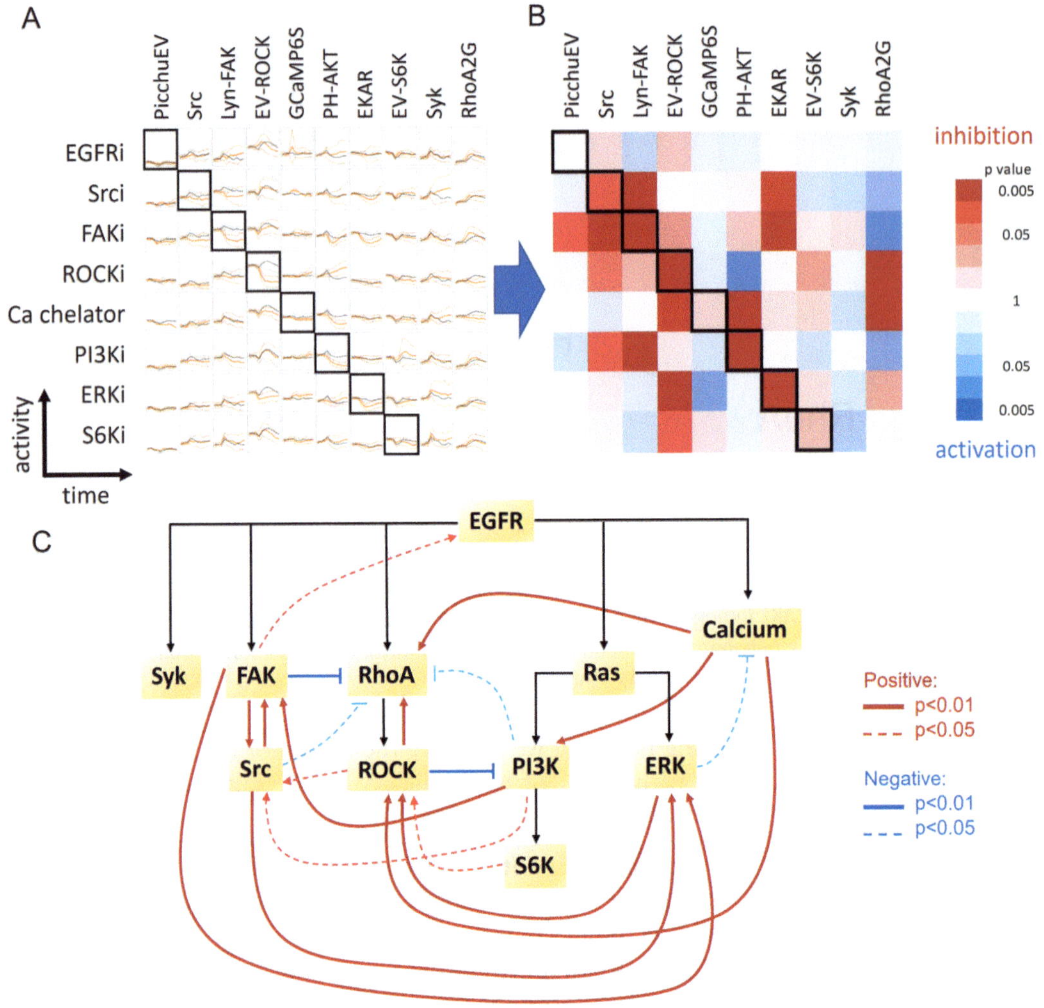

Fig. 4 Constructing the interaction map (**a**) Responses of biosensors to inhibitors are displayed as traces in orange, while the traces in gray indicate the responses to the vehicle control. (**b**) The matrix of inhibition displays how inhibitors affect the activities of biosensors. Red and blue colors indicate negative and positive effects, respectively, while the color scale indicates the statistical significance. Biosensors associated with the targets of inhibitors are indicated by black boxes. From the inhibition matrix, feedback loops are deduced by assigning positive (red) or negative (blue) interactions based on whether an inhibitor resulted in the inhibition or activation, respectively, of another node in the network

cells. However, it is important to consider that certain interactions may be indirect or influenced by unintended effects of the inhibitors and could vary depending on the cell type being studied. Nonetheless, these findings emphasize the intricate regulatory connections between nodes in the EGFR signaling network and demonstrate an effective approach to uncovering previously unknown interactions within signaling networks of specific cell types.

4 Notes

1. GFP and YFP are both imaged under the excitation laser for CFP (458 nm). In some publications, the YFP channel is labeled as the FRET channel, and the YFP/CFP ratio as the FRET/CFP ratio.
2. In order to optimize your transfection process, it is recommended to work with healthy cells that have undergone only a few passages. Regular maintenance of the cells is also important to prevent overcrowding. When it comes to transfection, it is best to aim for a cell density of 70–80%.
3. If the barcode signal is too low, increase the DNA amount during transfection. Additionally, make sure to use high-quality plasmid DNA that is free from phenol, sodium chloride, and endotoxins.
4. During focus adjustment or cell search, it is recommended to lower the laser power, minimize exposure time, and utilize RFP instead of BFP due to its susceptibility to quick bleaching.
5. Depending on the velocity of the motorized stage and the time interval between consecutive image frames, the number of positions can be adjusted to optimize the number of cells captured and permit ample time for both image acquisition and EGF administration.
6. If it proves challenging to encompass the entire cell within the chosen region, try incorporating a portion of the nucleus and plasma membrane. In addition, exclude apoptotic cells, debris, or impurities from the area of interest.
7. The thresholds for models 2 and 3 are both set to 0.9 by default. While a greater threshold typically results in more precise outcomes, it also has the tendency to recognize fewer cells.

Acknowledgments

The plasmids of PicchuEV, EV-ROCK, and EV-S6K were provided by Michiyuki Matsuda, to whom the authors extend their gratitude. This work was supported by funding from R01GM136711 (to C. H.H.), Cervical Cancer SPORE P50CA098252 Career Development Award (to J.M.Y.) and Pilot Project Award (to C.H.H.), the W. W. Smith Charitable Trust (#C1901, to C.H.H.) and the Sol Goldman Pancreatic Cancer Research Center (to C.H.H.). Additionally, the purchase of the Zeiss LSM 780 confocal microscope was made possible by NIH grants S10OD016374.

References

1. Greenwald EC, Mehta S, Zhang J (2018) Genetically encoded fluorescent biosensors illuminate the spatiotemporal regulation of signaling networks. Chem Rev 118:11707–11794. https://doi.org/10.1021/acs.chemrev.8b00333
2. Terai T, Campbell RE (2022) Barcodes, co-cultures, and deep learning take genetically encoded biosensor multiplexing to the nth degree. Mol Cell 82:239–240. https://doi.org/10.1016/j.molcel.2021.12.017
3. Yang J-M, Chi W-Y, Liang J et al (2021) Deciphering cell signaling networks with massively multiplexed biosensor barcoding. Cell 184:6193–6206.e14. https://doi.org/10.1016/j.cell.2021.11.005
4. Chi W-Y, Au G, Liang J et al (2022) Imaging and analysis for simultaneous tracking of fluorescent biosensors in barcoded cells. STAR Protoc 3:101611. https://doi.org/10.1016/j.xpro.2022.101611
5. Komatsu N, Aoki K, Yamada M et al (2011) Development of an optimized backbone of FRET biosensors for kinases and GTPases. Mol Biol Cell 22:4647–4656. https://doi.org/10.1091/mbc.E11-01-0072
6. Fritz RD, Letzelter M, Reimann A et al (2013) A versatile toolkit to produce sensitive FRET biosensors to visualize signaling in time and space. Sci Signal 6:rs12. https://doi.org/10.1126/scisignal.2004135
7. Seong J, Ouyang M, Kim T et al (2011) Detection of focal adhesion kinase activation at membrane microdomains by fluorescence resonance energy transfer. Nat Commun 2:406. https://doi.org/10.1038/ncomms1414
8. Xiang X, Sun J, Wu J et al (2011) A FRET-based biosensor for imaging SYK activities in living cells. Cell Mol Bioeng 4:670–677. https://doi.org/10.1007/s12195-011-0211-x
9. Várnai P, Balla T (1998) Visualization of phosphoinositides that bind pleckstrin homology domains: calcium- and agonist-induced dynamic changes and relationship to myo-[3H]inositol-labeled phosphoinositide pools. J Cell Biol 143:501–510. https://doi.org/10.1083/jcb.143.2.501
10. Harvey CD, Ehrhardt AG, Cellurale C et al (2008) A genetically encoded fluorescent sensor of ERK activity. Proc Natl Acad Sci USA 105:19264–19269. https://doi.org/10.1073/pnas.0804598105
11. Chen T-W, Wardill TJ, Sun Y et al (2013) Ultrasensitive fluorescent proteins for imaging neuronal activity. Nature 499:295–300. https://doi.org/10.1038/nature12354
12. Li C, Imanishi A, Komatsu N et al (2017) A FRET biosensor for ROCK based on a consensus substrate sequence identified by KISS technology. Cell Struct Funct 42:1–13. https://doi.org/10.1247/csf.16016
13. Ouyang M, Sun J, Chien S et al (2008) Determination of hierarchical relationship of Src and Rac at subcellular locations with FRET biosensors. Proc Natl Acad Sci USA 105:14353–14358. https://doi.org/10.1073/pnas.0807537105
14. Schindelin J, Arganda-Carreras I, Frise E et al (2012) Fiji: an open-source platform for biological-image analysis. Nat Methods 9:676–682. https://doi.org/10.1038/nmeth.2019
15. Schneider CA, Rasband WS, Eliceiri KW (2012) NIH image to ImageJ: 25 years of image analysis. Nat Methods 9:671–675. https://doi.org/10.1038/nmeth.2089
16. Chollet F (2018) Keras: the python deep learning library. Astrophysics Source Code Library. Available: https://ui.adsabs.harvard.edu/abs/2018ascl.soft06022C/abstract
17. Abadi M, Agarwal A, Barham P et al (2016) TensorFlow: large-scale machine learning on heterogeneous distributed systems. arXiv [cs.DC]. Available: http://arxiv.org/abs/1603.04467

Chapter 14

Deep Learning-Based Cell Tracking in Deforming Organs and Moving Animals

Chentao Wen

Abstract

Cell tracking is an essential step in extracting cellular signals from moving cells, which is vital for understanding the mechanisms underlying various biological functions and processes, particularly in organs such as the brain and heart. However, cells in living organisms often exhibit extensive and complex movements caused by organ deformation and whole-body motion. These movements pose a challenge in obtaining high-quality time-lapse cell images and tracking the intricate cell movements in the captured images. Recent advances in deep learning techniques provide powerful tools for detecting cells in low-quality images with densely packed cell populations, as well as estimating cell positions for cells undergoing large nonrigid movements. This chapter introduces the challenges of cell tracking in deforming organs and moving animals, outlines the solutions to these challenges, and presents a detailed protocol for data preparation, as well as for performing cell segmentation and tracking using the latest version of 3DeeCellTracker. This protocol is expected to enable researchers to gain deeper insights into organ dynamics and biological processes.

Key words Cell tracking, Deep learning, 3D time-lapse images, Deforming organs, Moving animals

1 Introduction

1.1 Tracking Cells in Deforming Organs and Moving Animals

To comprehend the mechanisms underlying various biological functions and processes in organs such as the brain and heart, scientists need to extract cellular signals from a group of cells within these organs. In living organisms, particularly animal organs, cells often exhibit movements caused by organ deformation and whole-body movement (e.g., in neurons of semi-constrained and freely moving worms, and cardiomyocytes cells in beating zebrafish hearts) [1–5]. Such movements pose substantial challenges in extracting cellular signals from in vivo images, as the regions for signal extraction in each cell can greatly change over time. To address this issue, cell tracking techniques are essential.

Traditionally, cell tracking techniques have been more extensively studied in developing organism [6], where images are

Christoph Wülfing and Robert F. Murphy (eds.), *Imaging Cell Signaling*, Methods in Molecular Biology, vol. 2800, https://doi.org/10.1007/978-1-0716-3834-7_14,

obtained from embryos or immature organs. In these images, cell movements are less pronounced, but cell division is common. Thus, the main challenge is detecting cell division events and establishing cell lineage over time. In contrast, cell tracking in mature organs, such as brains and hearts, presents different characteristics: the images are from mature organs with minimal cell divisions, but the organs can undergo substantial deformations. This article focuses on addressing the challenges of cell tracking in 3D images of deforming mature organs, where a deformable cell population with long-distance movements are the primary concern.

1.2 Techniques for Tracking Cells

Cell tracking is a specialized task within the broader field of multi-object tracking, a computer vision technique with applications in autonomous driving, robot navigation, and more [7]. Unlike general multi-object tracking, which typically involves analyzing 2D videos containing objects from various categories, cell tracking specifically focuses on tracking time-lapse images of cells obtained from microscopes [6].

Tracking cells generally involves two basic steps: extracting individual cells from images and estimating cell movements at different time points. A variety of techniques can be applied to both steps. Traditional, algorithm-based techniques are usually easier to understand and implement. For example, cell detection can be achieved using a Gaussian filter followed by thresholding [8], while cell tracking can be accomplished through nearest neighbor linking [6]. However, these methods may struggle to process images under challenging conditions, such as low image quality, complex cell shapes, densely packed cells, and long-distance cell movements in deforming organs.

In contrast, recent advanced techniques primarily rely on machine learning, particularly deep learning. While deep learning-based methods have limited explainability and can be considered "black boxes," they have become increasingly popular for cell detection and tracking in recent years (*see* Subheading 1.3). This is primarily due to their ability to handle more challenging conditions compared to traditional techniques. Nonetheless, traditional methods still play a significant role in pre- and post-processing to enhance the performance of deep learning-based techniques in cell tracking applications.

1.3 Tracking Cells with Deep Learning Techniques

Deep learning techniques refer to a type of machine learning method that employs artificial neural networks with multiple layers to learn to make predictions from training data. These techniques have been widely used in image and language processing tasks, often achieving better performance than non-deep learning methods [9, 10].

Deep learning techniques have also been widely employed in general multi-object tracking [7] and cell tracking [11]. These methods can learn how to detect and track cells by analyzing numerous examples, or training data, provided by users. This training data consists of cell images, along with the correct results of cell detection and tracking. Once trained, the model can make accurate predictions about cell detection and tracking on new, unseen images.

As previously mentioned, the first step in tracking cells is extracting individual cells from the images, a process known as cell segmentation. Various deep learning techniques, such as U-Net [12], StarDist [13], and CellPose [14], have been proposed and have demonstrated excellent performance on diverse datasets.

The second step in tracking cells, estimating cell movements, can also be performed using deep learning techniques. Depending on the scenario, a previous method for tracking cells in developing organisms [15] used deep learning techniques to estimate optical flow [16, 17], wherein a deep neural network is trained to directly predict cell movements by comparing raw images at two different time points. However, these methods often struggle to capture the spatial relationships of cells in the entire image, leading to difficulties in tracking cells with long-distance movements. In contrast, other deep learning models [18–20] match cells based on the cell coordinates extracted in the first step, cell segmentation. Since these models consider the spatial relationships of cells in the entire image, they are better suited for tracking cells that exhibit substantially larger movements in deforming organs.

1.4 Overview of the Cell Tracking Method Detailed in This Protocol

In Subheading 2, a detailed protocol is presented for the cell tracking method, which relies on the deep learning model StarDist [13] to extract individual cells and uses another deep network, FFN [18], to predict matchings between cells at different time points (Fig. 1). The predicted matching is improved by considering the coherence feature of cell movements in deforming organs, meaning that neighboring cells should have similar movements. By taking this coherence into account, the method can generate more accurate predictions of cell movements. The method has been tested on a variety of biological samples, including neurons in freely moving worms, cardiomyocytes cells in the beating heart of living zebrafish larvae, and cells in cultured 3D tumor spheroid, as reported in a previous paper [18]. The versatility of the method makes it suitable for a wide range of applications in cell tracking for various biological samples with substantial organ deformations.

The initial proposal of this cell tracking method was made in [18], with a detailed protocol provided for an older version (v0.4) in [21]. Subsequently, the program was updated to version v1.0, and three substantial improvements were made to enhance its performance and usability. These improvements include the following:

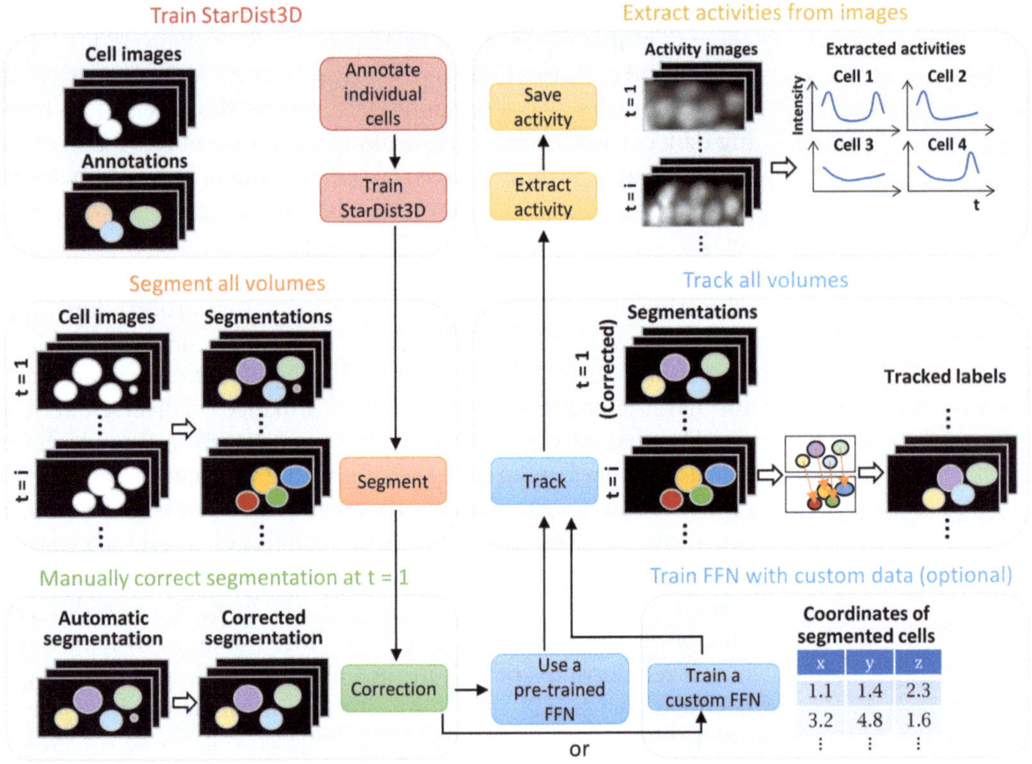

Fig. 1 Overview of the entire workflow of the cell tracking method, illustrating the main steps involved in the protocol

1. Previously, cell segmentation was performed with 3D U-Net [12], a semantic segmentation [22] method that classifies each pixel as a cell or non-cell region. As a result, the predicted cell regions required additional processing, such as watershed [23], to separate them into individual cells. To simplify the cell segmentation step, a new API was made, which allows users to employ the StarDist3D method [13] to directly generate instance segmentation [22] results, that is, cell regions separated into individual cells (Fig. 1). The improved workflow with StarDist requires fewer user-specified parameters and is much more convenient to use than the previous version.
2. The tracking system was optimized by normalizing the cell coordinates extracted in **step 1** before performing cell matching and estimating movements. This update makes the default tracking parameters work well on most image datasets and simplifies the fine-tuning of the parameters compared to the previous version.
3. Users now can train their own FFN to achieve more accurate tracking of custom data (Fig. 1).

1.5 Advantages and Limitations of the Cell Tracking Method

The method is designed to track cells in deforming organs with large nonrigid deformations. It can track cells under such conditions even when the movements are considerably larger than the cell sizes, the image quality is suboptimal, and the cells are densely attached with each other. On the other hand, the method does not include algorithms to handle cell divisions and cell death, as it is primarily designed to track cell movements in deforming organs, where cell divisions are rare. Therefore, it cannot reconstruct the cell lineage in developing organisms, where cell divisions and death are more common. Users seeking to track cells in such scenarios may need to explore alternative methods or modify the method to incorporate additional features that can handle cell divisions and death. Moreover, the method is designed to track cells in deforming organs, wherein user-specified cells in the initial volume stay within the field of view throughout the time-lapse imaging process. Consequently, it is not suitable for tracking rapidly migrating cells that quickly exit the field of view, such as red blood cells. Additionally, the method can only track cells in 3D time-lapse images, not in 2D images.

2 Methods

2.1 Prepare the Computational Environment

1. Check GPU compatibility and install the required software components/libraries:

   Ensure your computer has a compatible NVIDIA GPU (check at https://developer.nvidia.com/cuda-gpus). Then, download and install the NVIDIA CUDA Toolkit and cuDNN library following their respective installation guides (https://docs.nvidia.com/cuda/cuda-quick-start-guide/index.html and https://docs.nvidia.com/deeplearning/cudnn/index.html). Please note that the versions of CUDA and cuDNN should be compatible with the TensorFlow version (*see* **step 3**).

2. Download and install Miniconda from the official website: https://docs.anaconda.com/free/miniconda/index.html. Choose the appropriate version for your operating system (OS) and follow the installation instructions.

3. Open your computer's command prompt (Windows) or terminal (Linux) to install TensorFlow in a conda environment. Create a new conda environment with Python by typing:

   ```
   $ conda create --name tf_environment python=3.8 pip
   ```

   Press Enter to create the environment. Activate the new environment by typing:

$ conda activate tf_environment

Now, carefully read and follow the TensorFlow installation guide to set up TensorFlow in the newly created environment: https://www.tensorflow.org/install. Please pay attention to the compatibility requirements between TensorFlow, CUDA, and cuDNN. During testing, TensorFlow 2.5.0 or TensorFlow 2.6.0 worked on the Ubuntu 20.04 system with an Nvidia GeForce RTX 3080 Ti. However, depending on your OS and GPU, you may need to choose a different version.

4. With the "tf_environment" activated, type the following command to install 3DeeCellTracker using pip, Python's package installer:

   $ pip install 3DeeCellTracker==1.0.0

   Press Enter, and the 3DeeCellTracker program will be installed on your computer.

2.2 Download the Jupyter Notebooks and the Demo Data

1. All of the notebooks described in this protocol are stored in the following GitHub repository: https://github.com/WenChentao/3DeeCellTracker/blob/master/Examples/use_stardist/

   Here is the list of the notebooks described in this protocol:

   train_stardist.ipynb

   track_stardist_single_mode.ipynb

   track_stardist_ensemble_mode.ipynb

   train_ffn_with_coordinates_csv.ipynb

   train_ffn_with_label_image.ipynb

   Users can download the notebooks to their local computer by clicking the "Download" button on the GitHub page or by cloning the repository.
2. The program in the notebooks allows users to analyze their own data. In addition, pre-trained models and example data are provided to demonstrate how to use the program. These data and models are available at https://osf.io/pgr95/ (except for worm3_points_t1.csv).

   Here is the list of the data and models:

   worm1_stardist_training_data.zip

   worm1.zip

   worm4.zip

   stardist_models.zip

   ffn_models.zip

   worm3_points_t1.csv (in the GitHub repository mentioned above)

2.3 Cell Segmentation

1. Prepare training data for cell segmentation

   To use StarDist for cell segmentation, users first need to train the model. The training data consists of 3D raw images of cells and their corresponding 3D annotations of label images, which indicate the cell IDs for each pixel. Save the raw and label images as TIFF image stacks in two separate folders, ensuring that the corresponding images have identical filenames so the program can recognize their relationship. For training, provide at least one 3D raw image and its corresponding 3D label image as training data (Fig. 2). Create label annotations by generating labels using an automatic method (e.g., applying a thresholding following by watershed in Fiji) and then correct any mistakes in the labels using software like ITK-SNAP. For a quick trial, use the demo data "worm1_stardist_training_data."

2. Train the StarDist model

   With the training data prepared, train the StarDist model using the provided training program in the notebook: train_-stardist.ipynb. To start the Jupyter Notebook server, activate the "tf_environment" and run the following command:

   ```
   $ jupyter notebook
   ```

   This opens the Jupyter Notebook server in your browser. Once you have opened the downloaded notebook, execute each code block sequentially. Keep in mind that you may need to modify parameters in some code blocks according to the instructions provided in the notebook. Here's a brief overview of how to use this notebook for training StarDist:

   In the notebook, start by specifying the paths of the raw and corresponding label images to load the training data. Next, set a filename for the trained StarDist model. After that, set the number of training epochs; the default value of 500 is suitable for the demo data, but users can increase it to potentially improve segmentation results. Once the training is complete, the model will be automatically saved to the "stardist_models" folder with the specified filename, allowing users to load the model later for cell segmentation in time-lapse images.

3. Perform cell segmentation with trained StarDist model

   After training the StarDist model, users can perform cell segmentation using the trained model within the notebook track_stardist_single_mode.ipynb or track_stardist_ensemble_mode.ipynb. Users have the option to perform segmentation and tracking on their custom images with custom StarDist and FFN models trained on their own images. Or they can use the demo datasets (worm1.zip and worm4.zip) along with the pre-trained models (stardist_models.zip and ffn_models.zip).

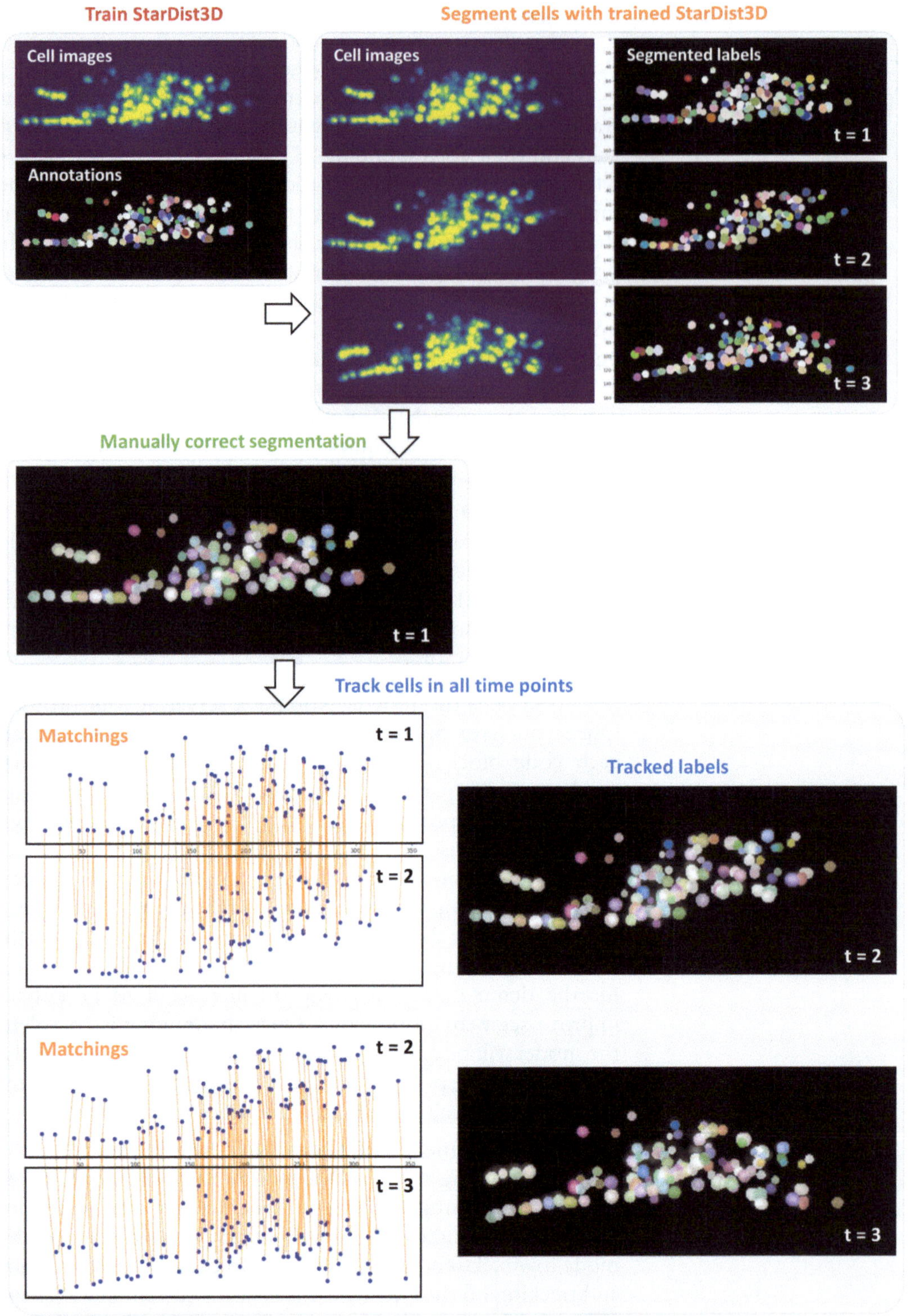

Fig. 2 Examples of cell segmentation and tracking results at various stages of the protocol. The colors in the annotations/segmentations are added to show the different cell IDs. The blue dots in the matching panel indicate the positions of the segmented cells in two different volumes, whereas the orange lines indicate the predicted movements from the first volume to the second volume. The raw time-lapse 3D images depict head neurons from a straightened, freely moving worm [2]. For illustrative purposes, all results have been converted to 2D space using maximum intensity projection

To prepare the custom images to be segmented and tracked, users must first convert raw time-lapse 3D image data into 2D TIFF images, which include T time points and Z slices. Each image file should be named with timing (t) and slice number (z) information, such as "image_t0001_z0001.tif." Ensure that the timing values follow a continuous sequence, like t0001, t0002, t0003, and so on, and maintain a uniform format for the filenames. In the same way, consistently represent z values and maintain their range throughout the entire time.

After preparing and saving the raw images in a designated folder, users need to configure several paths for the program. First, provide the trained StarDist model's name so the program can load it. Next, set a folder path where the program will store all segmentation and tracking results. Finally, specify the paths to the raw images. For example, the path to the images can be formatted as "./animal1/data/**t%04d**.tif," where "t%04d" represents a "t" followed by a four-digit number.

Once the paths have been configured, execute the code blocks in Section B. The program will automatically create a segmentation of the 3D image at volume #1 (Fig. 2) and save the result (label images) into the "auto_seg1" folder, which the user must manually review and correct. Additionally, the program will segment all volumes in the raw image folder sequentially (Fig. 2) and save the segmentation results: cell coordinates, and images of the probability map of the cell regions, which will be used in the subsequent tracking process.

4. Manually correct segmentation results at vol #1

   The automatic segmentation process may contain a few mistakes, such as cells classified as background, artifacts classified as cells, or incorrect separations of multiple attached cells, and so on. These mistakes require users to manually correct them (Fig. 2), but they only need to correct the segmentation for the first volume. The tracking methods will use the corrected segmentation of cells as seeds and predict their new positions in subsequent volumes.

   The manual correction of segmentation results can be done using third-party software such as ITK-SNAP. After completing the correction, users need to save the corrected segmentation as 2D TIFF images in a separate folder.

2.4 Cell Tracking

In this section, we discuss cell tracking. After performing segmentation on all volumes and correcting the first volume segmentation, users can begin cell tracking within the same notebook, i.e., track_stardist_single_mode.ipynb or track_stardist_ensemble_mode.ipynb. First, users need to load the manually corrected segmentation, and the program will carry out several processing steps on the loaded segmentation, which is required for cell tracking.

1. To load the corrected segmentation, users must set the path to the corrected segmentation in volume 1. Additionally, users need to set the voxel size in the x, y, and z axes, which will help the program transform the cell coordinates from their positions in the 3D image to their positions in real physical space. Users also need to set a parameter called the interpolation factor so the program can interpolate the 3D cell regions along the z-axis. For example, if a cell region originally contains three slices and the interpolation factor is 10, the interpolated cell region will contain 30 slices. This process helps the program generate smoother cell boundaries even if the resolution along the z-axis is much lower than in the x-y plane. As a result, when cells move along the z-axis in small steps, such as 0.1 slice, the program generates tracking results with more accurate cell boundaries. In addition to interpolation, the program also corrects mistakes that occur when two distant cells are mistakenly labeled as the same one during manual correction.
2. Once users have prepared the segmentation, it is time to track cells across time. Users can utilize the provided pre-trained FFN model to track cells within these images, which is designed to function effectively in most scenarios. Alternatively, users have the option to train a customized FFN model with their own data (*see* Subheading 2.5).
3. After setting the path and filename for the FFN model correctly, the program loads the model. Users can test the matching outcomes between two sets of cells segmented at different times using the FFN model. The timing 1 and timing 2 can be freely adjusted to inspect the robustness of the matching accuracy.
4. If the FFN matching achieves an accuracy of approximately 90% or higher, users can proceed to test the predicted cell movements based on the FFN matching and PR-GLS [24], which filters incoherent cell matchings to improve tracking accuracy. The default values of the PR-GLS parameters beta and lambda typically work well, but users can modify them to enhance the results.
5. Once the parameters have been well configured and satisfactory tracking results are obtained during the tests, users can proceed to sequentially track cells across all volumes. The program predicts the updated center positions of cells defined in the confirmed segmentation in volume #1 for each subsequent volume. Two tracking modes are applicable in different scenarios: Single mode and ensemble mode. Users can try the single mode with track_stardist_single_mode.ipynb, which predicts cell positions from time t to $t + 1$ sequentially. This mode is quick but could be less accurate in challenging datasets,

such as those with large cell movements or poor image quality. If the single mode does not provide satisfactory results, users can attempt the ensemble mode with track_stardist_ensemble_mode.ipynb. The ensemble mode predicts the cell positions at time t as the average of multiple predictions from different timings preceding t, resulting in increased accuracy but at the expense of longer processing time.

6. The predicted cell coordinates and tracked label images will be saved in the "track_results" folder automatically for further analyses. Additionally, the figures displaying the tracking process with points and lines, and the merged image of raw and label images will also be saved for diagnostic purposes (Fig. 2). To bypass specific volumes with poor/unusable image quality, users can set a parameter skip_volumes = [t1, t2, t3, ...], enabling the tracking program to skip these volumes.

2.5 Train FFN Model (Optional)

While the provided pre-trained FFN works across different datasets, training a custom FFN model on users' own data may further improve the tracking accuracy by better capturing the specific features of cell positions (Fig. 1).

There are two methods for training an FFN model. The first method uses segmented label images and can be done using the "train_ffn_with_label_image.ipynb" notebook. Users provide the voxel size and the path to a directory containing a single volume of segmented label images. The program extracts the coordinates from the images and transforms them into physical coordinates for training the FFN model.

The second method involves providing a .csv file of 3D coordinates of a point set. Users can use the "train_ffn_with_coordinates_csv.ipynb" notebook to start the training process by specifying the path to the .csv file. It is important to note that the 3D coordinates in the .csv file should represent the cell positions in real physical space.

In either case, users should specify a name for the model. This is necessary because the trained model will be saved using the specified name, and the same name will be needed to load the model during the tracking process. Additionally, users should set the number of training epochs, with a default value of 100 being a good starting point, but users can increase it to potentially improve tracking accuracy.

2.6 Extract Cellular Activity

After tracking cells across all time points, users can extract cell activities from the tracked cell regions at each time point in the same Jupyter notebook, i.e., track_stardist_single_mode.ipynb or track_stardist_ensemble_mode.ipynb (Fig. 1). To accomplish this, users should specify the path to the raw images containing cell activities. It is important to note that these images may either be the same as the ones previously used for segmentation and tracking, or they could be other simultaneously captured cell images from a different channel.

The extracted activities can be visualized using curves representing each activity. To export the activities to a .csv file, users must provide a filename. The program will save the results with this filename in the folder specified by users for storing all segmentation and tracking results (as described in Subheading 2.3).

Acknowledgments

This work was supported by the Special Postdoctoral Researchers Program in RIKEN (to C.W.).

References

1. Toyoshima Y, Tokunaga T, Hirose O et al (2016) Accurate automatic detection of densely distributed cell nuclei in 3D space. PLoS Comput Biol 12(6):e1004970
2. Nguyen JP, Shipley FB, Linder AN et al (2016) Whole-brain calcium imaging with cellular resolution in freely behaving Caenorhabditis elegans. Proc Natl Acad Sci USA 113:E1074–E1081
3. Venkatachalam V, Ji N, Wang X et al (2016) Pan-neuronal imaging in roaming Caenorhabditis elegans. Proc Natl Acad Sci USA 113: E1082–E1088
4. Mickoleit M, Schmid B, Weber M et al (2014) High-resolution reconstruction of the beating zebrafish heart. Nat Methods 11:919–922
5. Voleti V, Patel KB, Li W et al (2019) Real-time volumetric microscopy of in vivo dynamics and large-scale samples with SCAPE 2.0. Nat Methods 16:1054–1062
6. Ulman V, Maška M, Magnusson KEG et al (2017) An objective comparison of cell-tracking algorithms. Nat Methods 14:1141–1152
7. Ciaparrone G, Luque Sánchez F, Tabik S et al (2020) Deep learning in video multi-object tracking: a survey. Neurocomputing 381:61–88
8. Vicar T, Balvan J, Jaros J et al (2019) Cell segmentation methods for label-free contrast microscopy: review and comprehensive comparison. BMC Bioinf 20:360
9. Russakovsky O, Deng J, Su H et al (2015) ImageNet large scale visual recognition challenge. Int J Comput Vis 115:211–252
10. Brown T, Mann B, Ryder N et al (2020) Language models are few-shot learners. In: Advances in neural information processing systems. Curran Associates, Inc., pp 1877–1901
11. Moen E, Bannon D, Kudo T et al (2019) Deep learning for cellular image analysis. Nat Methods 16:1233–1246
12. Çiçek Ö, Abdulkadir A, Lienkamp SS et al (2016) 3D U-net: learning dense volumetric segmentation from sparse annotation. In: Medical image computing and computer-assisted intervention—MICCAI. Springer, Cham, pp 424–432
13. Weigert M, Schmidt U, Haase R et al (2020) Star-convex polyhedra for 3D object detection and segmentation in microscopy. In: 2020 IEEE winter conference on applications of computer vision (WACV). IEEE, Snowmass Village, pp 3655–3662
14. Stringer C, Wang T, Michaelos M, Pachitariu M (2021) Cellpose: a generalist algorithm for cellular segmentation. Nat Methods 18:100–106
15. Sugawara K, Çevrim Ç, Averof M (2022) Tracking cell lineages in 3D by incremental deep learning. elife 11:e69380
16. Dosovitskiy A, Fischer P, Ilg E et al (2015) FlowNet: learning optical flow with convolutional networks. In: 2015 IEEE international conference on computer vision (ICCV), pp 2758–2766
17. Ilg E, Mayer N, Saikia T et al (2017) FlowNet 2.0: evolution of optical flow estimation with deep networks. In: Proceedings of the IEEE conference on computer vision and pattern recognition (CVPR), pp 2462–2470
18. Wen C, Miura T, Voleti V et al (2021) 3DeeCellTracker, a deep learning-based pipeline for segmenting and tracking cells in 3D time lapse images. elife 10:e59187

19. Yu X, Creamer MS, Randi F et al (2021) Fast deep neural correspondence for tracking and identifying neurons in C. elegans using semi-synthetic training. elife 10:e66410
20. Wu Y, Wu S, Wang X et al (2022) Rapid detection and recognition of whole brain activity in a freely behaving Caenorhabditis elegans. PLoS Comput Biol 18:e1010594
21. Wen C, Kimura KD (2022) Tracking moving cells in 3D time lapse images using 3DeeCell-Tracker. Bio Protoc 12:e4319
22. Minaee S, Boykov Y, Porikli F et al (2022) Image segmentation using deep learning: a survey. IEEE Trans Pattern Anal Mach Intell 44:3523–3542
23. Beucher S, Meyer F (1993) The morphological approach to segmentation: the watershed transformation. In: Dougherty ER (ed) Mathematical morphology in image processing. Marcel Dekker, New York, pp 433–481
24. Ma J, Zhao J, Yuille AL (2016) Non-Rigid point set registration by preserving global and local structures. IEEE Trans Image Process 25: 53–64

Chapter 15

Using Dimensionality Reduction to Visualize Phenotypic Changes in High-Throughput Microscopy

Alex X. Lu and Alan M. Moses

Abstract

High-throughput microscopy has enabled screening of cell phenotypes at unprecedented scale. Systematic identification of cell phenotype changes (such as cell morphology and protein localization changes) is a major analysis goal. Because cell phenotypes are high-dimensional, unbiased approaches to detect and visualize the changes in phenotypes are still needed. Here, we suggest that changes in cellular phenotype can be visualized in reduced dimensionality representations of the image feature space. We describe a freely available analysis pipeline to visualize changes in protein localization in feature spaces obtained from deep learning. As an example, we use the pipeline to identify changes in subcellular localization after the yeast GFP collection was treated with hydroxyurea.

Key words Subcellular localization, Deep Learning, Image analysis, Proteins

1 Introduction

Automated image analysis of microscopy images is a key technique to enable quantitative, objective, and data-rich cell biology [1]. In the past decade, deep learning has revolutionized many aspects of biological image analysis [2, 3]. For example, cell-segmentation for some types of cells is now an automated procedure that in many cases can be done reasonably accurately without any user intervention (no training data or adjusting parameters [4–6]). Image features (numerical transformations of images to enable computational analysis) are also now being developed through deep learning. This means expert knowledge for each dataset is not required, and the learned features often show improved performance because they can be more sensitive to the nuances of the data than experts working from first principles. Of particular interest are unsupervised deep learning strategies where labelled data is not needed for the neural network to learn descriptive features. Self-supervised learning [7], a type of unsupervised learning, overcomes

Christoph Wülfing and Robert F. Murphy (eds.), *Imaging Cell Signaling*, Methods in Molecular Biology, vol. 2800,
https://doi.org/10.1007/978-1-0716-3834-7_15,

the labor and bias introduced by needing to label extensive amounts of training data by instead training neural networks on "proxy tasks."

These proxy tasks do not need to produce directly useful outputs, but should be fully autonomous (so no expert labeling of data is required) and teach the model transferable features useful to the actual task of interest. Several such strategies have been proposed for microscopy images in the past few years [8–11].

Quantification of similarity or differences in cellular phenotype from images is among the most important image analysis tasks for high-throughput screens [12]. In some cases, images can be compared by experts, but this becomes infeasible as the numbers of images grow or the changes are too subtle to be reliably determined visually. Change detection for cellular phenotype is a challenging task to automate because of unimportant visual differences in cells due to technical variation such as cell orientation, illumination, magnification, and other microscope settings, as a well as biological variation such as cell density, normal cell shape variation, and cell health, whose importance depends on the biological question asked in the experiment. It is therefore important to distinguish specific changes in a marker or phenotype of interest from global changes (either biological or technical) that affect cellular appearance that are due to the experimental conditions.

One approach to automatic identification of cell phenotype changes is to train classifiers that recognize discrete cell phenotypes and apply them to find differences in the control and treatment conditions [12–14]. While powerful, this type of approach can be sensitive to covariate shifts [15], because fitting classifiers to ignore unimportant technical and biological variation has proven to be very difficult. New classifiers may need to be trained for each treatment condition where changes in cell phenotype are to be identified [16]. Several approaches have been developed to define phenotypes without defining training data. These include outlier detection approaches that classify phenotypes that differ too much from the wild-type [14, 17], projection of phenotypes onto predefined axes of interest [18], or application of clustering methods to automatically define the phenotype classes [19].

Alternatively, change detection can be conducted directly in the high-dimensional feature space, which has the advantage of not requiring prior knowledge of the phenotype classes of interest. However, performing change detection in the feature space creates new challenges because it is not obvious when changes in features become significant as changes in phenotype. To address this, we have developed a statistical method that constructs expectations to allow us to understand the significance of differences in features between experiments [20, 21]. Another major challenge is to visualize and explore the changes in cell phenotype. Because cell phenotypes are high-dimensional (they include the cell shape, the shape, size, texture and number of organelles [22], cell-cycle state [23], cell health [24], etc.) and are represented in high-dimensional features spaces, there are astronomically many different "ways" the phenotype can change. However, the particular way that the cell changes carries important biological and functional information, with changes along different combinatorial axes of phenotype potentially pointing to different underlying biological changes [21]; indeed, this is the basis for many applications of phenotype screening [12].

Inspired by visualization approaches now routinely used for single-cell sequencing analysis [25], we propose to visualize cell phenotype changes as arrows in a reduced-dimensional view of the feature space. Here, we describe an end-to-end pipeline to visualize changes in protein localization in feature spaces obtained from deep learning. This pipeline includes steps such as cell segmentation and feature extraction, which are now routinely performed using deep learning methods. As an illustrative example, we apply these visualization tools to a proteome-scale screen of the yeast-GFP collection [26] treated with hydroxyurea (HU).

2 Materials

1. Images from a microscopy screen in tif, PNG, or other open format (*see* **Note 1**)
2. Server/workstation with at least 32GB of RAM and 12GB GPU and 1 TB of storage space (*see* **Note 2**)
3. Python installation including numpy, PIL, matplotlib, skimage, sklearn, umap, keras, tensorflow (*see* **Note 3**)
4. Access to codes on Github (*see* **Note 4**)

3 Methods

3.1 Overview

Steps 1–3 represent standard image analysis operations that can be automated using deep learning. These steps can be thought of as the "step up" or preparation. **Steps 4** and **5** show how to visualize changes in morphology and subcellular localization in the image feature space.

All the codes needed to reproduce the analysis presented here are available at https://github.com/alexxijielu/change_detection_book_chapter/. Each subsection below corresponds to one of the steps marked in the Jupyter notebook included in that repository.

3.2 Segmenting Images into Single-Cell Crops

1. Typical single cell high-throughput microscopy images contain multiple cells in random locations and orientations, so the first step in our analysis will be segmenting these images into single cell crops. We will use YeastSpotter [5] to do this since it is a general purpose yeast cell-finding method that will work on many kinds of modalities (*see* **Note 5**).
2. Automated microscopes collect multichannel images, where each channel represents a stain, fluorescent tag, or imaging modality (such as brightfield, DIC, etc.). We will apply YeastSpotter to the cytoplasm channel, which in this screen is the RFP.
3. For automated analysis of high-throughput screen data, the segmentation method should be installed locally, as this process can take hours or days depending on the number of cells in the image collection.
4. YeastSpotter will produce segmentation masks where each cell is assigned its own unique integer. The individually numbered cells should then be saved as individual cell crop files.
5. We will rescale them and use 64×64 pixels as a standard size (*see* **Note 6**). This is important for deep learning methods that typically expect all input images to be the same size. Figure 1a shows an example of a raw image, and Fig. 1b shows the segmentation "mask" obtained from YeastSpotter. Figure 1c shows examples of individual cell "crops" that have been saved.

3.3 Create Features for This Dataset Using Self-Supervised Deep Learning on the Single-Cell Crops

To make the single cell crops easier to analyze, we convert them to "feature vectors" by computing numerical features that summarize the important features of the cell crop (*see* **Note 7**). Rather than designing these numerical features ourselves, we use a self-supervised method called paired cell inpainting [8]. This method trains a neural network on a "coloring book by example" problem: given an image of what protein expression looks like in one cell in an image, the model has to predict what protein expression will look like in a second cell from the same image.

3.4 Compute Protein-Level Features by Averaging Features Extracted by the Trained Deep Learning Model

1. Having trained a paired cell inpainting model, our next step is to use this model to extract features measuring protein localization in single yeast cells.
2. To do this, we will need to prune the model architecture after loading the weights for this model. The paired cell inpainting model is trained to output an image of predicted protein localization, but this image isn't directly useful for our goal of representing cells as features. Instead, we want the information that the model has learned to extract from the input images in order to solve the problem of generating this predicted output—this is our feature representation. We will need to prune

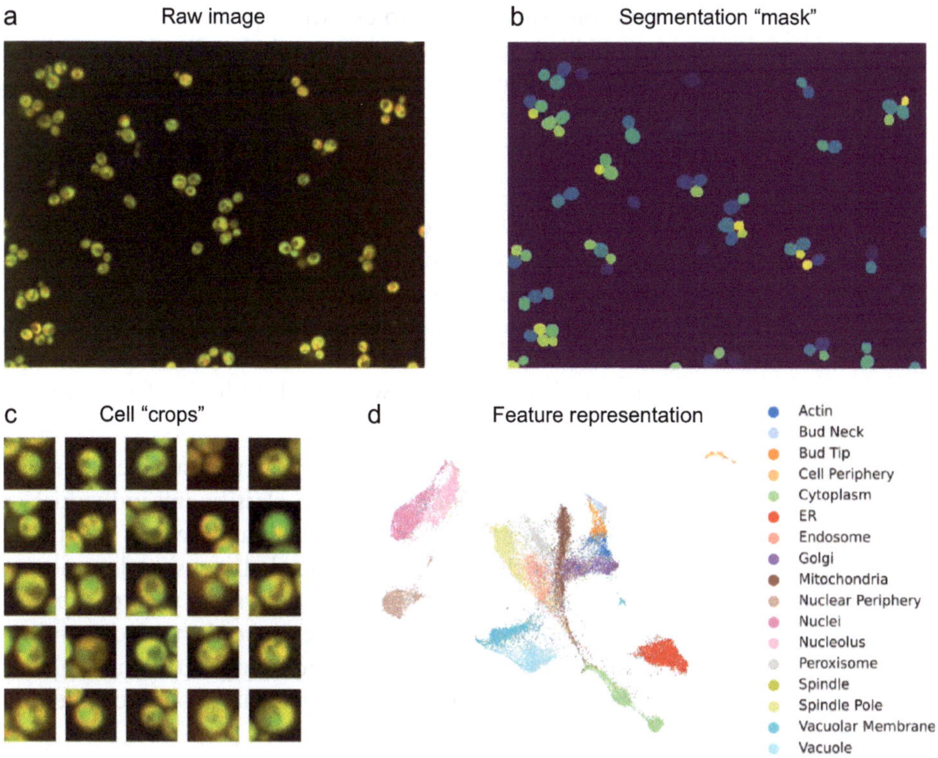

Fig. 1 Overview of standard automated image processing steps. (**a**) shows an example of a raw confocal microscope image. Here, one protein is tagged with GFP (in this case a cytoplasmic protein), and the cell is visible because it expresses a cytoplasmic RFP. (**b**) From the segmentation mask produced by YeastSpotter, each identified cell is shown as a different color. (**c**) based on the segmentation mask, "crops" are created where each one is centered around a single cell. (**d**) Each point indicates a single cell crop from a single-cell subcellular localization benchmark displayed in a 2D UMAP representation of image features automatically learned from our data using a self-supervised method (paired-cell inpainting). Colors represent the labels assigned to these cells by visual inspection [13]

away the decoder layers that generate the image and instead extract these features from an intermediate layer of an encoder in the paired cell inpainting model. Figure 1d shows image crops in this feature space as a UMAP [27] where the image crops are taken from a reference dataset of single-cell yeast images labelled through visual inspection [13].

3. To summarize an image or group of images using the learned features, we will simply average the feature representations of each identified cell in the image (*see* **Note 8**) to create a single feature vector for that image.

3.5 Visualizing Global Changes in Cellular Morphology

1. A first data analysis goal is to quantify systematic changes in cellular morphology (and other factors) in response to environmental perturbation. For example, consider a drug that causes vacuoles in yeast to fragment from a large single vacuole to

multiple smaller vacuoles. In a proteome-wide screen, we would likely find that every vacuole-localized protein has quantitatively changed localization under treatment with such a drug. However, we think of these changes as "markers" of vacuolar morphology, not protein-specific localization changes.

2. Hence, it is typical to consider a number of proteins (or stains) that localize to different components of the cell as "markers" that are then used to quantify changes in morphology. In principle, morphological changes should be easy to quantify in our yeast screen because we have imaged the localization of thousands of proteins: we have hundreds of markers for each cellular component.
3. Our goal here is to introduce a way to visualize changes in cell phenotypes, without having to classify any individual cells as "changed" or "not" or classify cells as "having" one phenotype or another. Because the reduced dimensionality view of the feature space (e.g., Fig. 1d) allows us to represent a large number of cellular phenotypes in only two dimensions, we reasoned that we could visualize changes in cellular phenotypes by adding arrows that indicate the direction of change on these plots.
4. Let's first visualize systematic changes in morphology. **Steps 1–3** have created feature representations for each GFP-tagged protein under control and treatment with the DNA damaging drug Hydroxyurea. We will visualize the changes in the feature space by drawing arrows between the control and treatment representation. The arrows represent the average condition-induced change in the representation space for proteins in the vicinity of the arrow (Fig. 2). In this case, we have created a 20×20 grid over the UMAP [27] space and plotted the average change for any box that had at least 20 proteins in it (*see* **Note 9**). According to this visualization, several regions of the feature space show concerted, systematic changes between the control and hydroxyurea treatment.
5. To interpret the changes in the feature space, we colored the individual protein feature representations in the control experiment by their original visual inspection of their localization from images [28]. We notice strong changes in the nucleus region (gray) suggesting changes in nuclear shape due to the drug, perhaps consistent with its action as a DNA damaging agent. We also notice changes in the cytoplasm (red) region of the feature space where mitochondria (purple) and vacuole (light blue) overlap whose interpretation is less clear. Some components such as the nucleolus and cell periphery don't seem to change much at all.

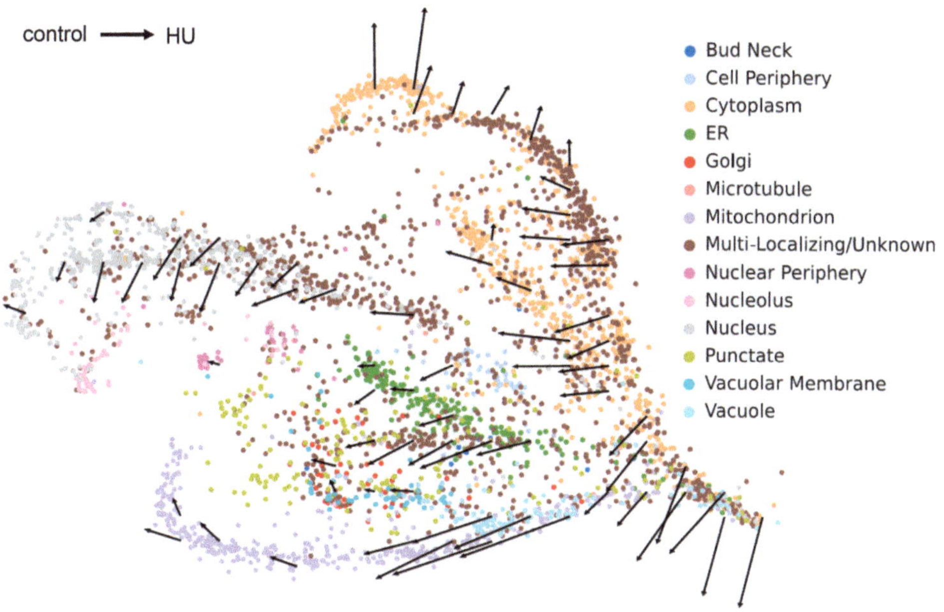

Fig. 2 Global morphology changes in the feature space. Each point represents a GFP-tagged yeast protein, where the single cell representations have been averaged to create a protein representation. These have been reduced to 2D and plotted. Arrows represent the change under HU treatment of the proteins nearby the beginning of the arrow in the representation space. Colors represent the localization class initially assigned to the GFP-tagged strains by experts looking at images taken under standard growth conditions

6. It is important to note that although we have interpreted these global changes as changes in cellular morphology, some of these global changes could be due to technical factors. Changes in the microscopy, illumination, etc. are also expected to create subtle global effects that may lead to shifts in the feature space.

3.6 Identifying and Visualizing Protein-Specific Changes in Subcellular Localization

1. Cells often respond to their environment by changing the localization of proteins, so identifying proteins that move to a different compartment in a cell in response to a perturbation can help characterize a cell's regulatory response.
2. Figure 3 shows an example pair of microscopy images from the CYCLoPS database [26]. Here, yeast cells have been treated with a DNA damaging drug (hydroxyurea, HU). The image shows confocal micrographs of a yeast strain expressing Msn2 tagged with GFP and a cytoplasmic RFP. It is visually obvious that the protein localization has changed upon drug treatment: while under the control conditions, most cells show GFP in the cytoplasm, upon trug treatment, most cells show nuclear localization (one bright circular GFP pattern in each cell).
3. A major data analysis goal is to systematically identify the proteins that change localization. It is important to note that, although in the example the protein changes from cytoplasm to

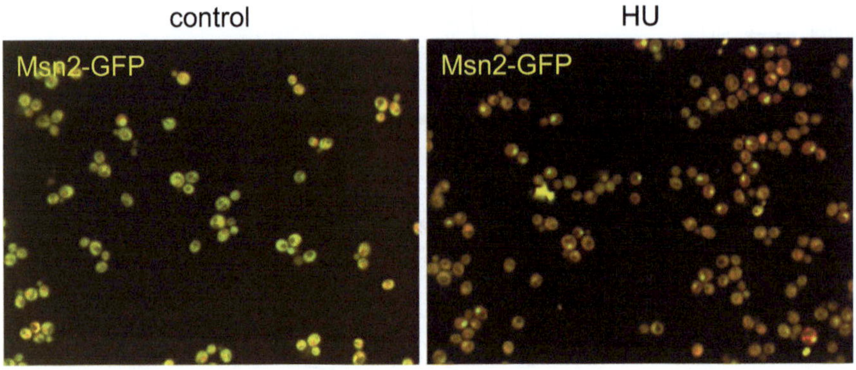

Fig. 3 Example of a change in localization pattern. GFP-tagged Msn2 ("Msn2-GFP") is mostly cytoplasmic under standard conditions ("control") and is mostly present in the cell nuclei after treatment with hydroxyurea ("HU")

nucleus, we don't want to specify in advance what the localization change will be. In principle, we might find that under drug treatment a protein localizes in a new pattern that we have never seen before: this would represent a new biological discovery, and we don't want to miss it!

4. As we have seen in Fig. 2, cells often exhibit systematic changes in morphology (and other factors), but identifying these is not the goal here: we'd like to find proteins that change in their specific subcellular pattern, as opposed to proteins that remain in the vacuole but look different only due to systematic differences in vacuole morphology and distribution.

5. Our solution to this problem is to employ a simple nearest-neighbor strategy [20, 21]: for each protein, we will compare its change to the 50 most similarly localized neighbors in the control. If the change differs drastically from its neighbors, we will consider that protein to actually be changing in localization. Otherwise, if the change is similar to that of its neighbors, even if the change is very large in feature space, we will consider this change to be a systematic one for most proteins localized to that compartment and dampen it.

6. To implement this concept, we calculate a z-score for each feature, of the protein compared to an expectation using the mean and standard deviation of its 50 nearest neighbors. Note that this analysis is done one feature at a time, so we have a Z-score for each dimension of the feature representation: we refer to these as change detection vectors.

7. Analyzing the change detection can be as simple as ranking the change detection vectors by magnitude—this produces a ranked list of putative protein localization changes. However, we have found it useful to visualize some of these changes.

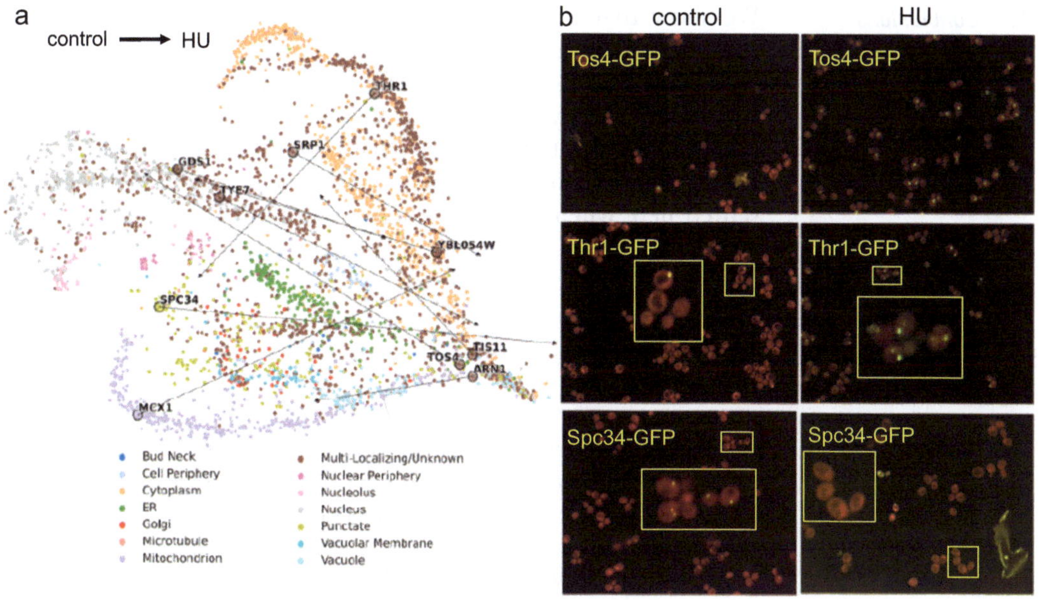

Fig. 4 Visualization of protein subcellular localization changes. (**a**) shows the top ten changes as identified by the k-nn change detection algorithm as arrows in the reduced dimensional image feature space. Arrows crossing from one part of the feature space to another indicate changes from one subcellular compartment to another. Points represent the locations of proteins in the control experiment, and the starting positions of the top ten changes are outlined with circles. The colors represent the labels given to each protein after visual inspection of the control condition. (**b**) raw microscope images for some of the changing proteins. Proteins are indicated on each image in control and HU conditions. Large yellow boxes show magnified views of the cells indicated in the small yellow boxes

8. Figure 4a shows the top ten changes ranked by the change detection algorithm, which controls for these systematic shifts. The arrows show corrected changes between the control and hydroxyurea treatment, and this visualization allows us to interpret the subcellular pattern of change, without having to say what the possible changes were in advance (as is be needed for an approach based on classification, e.g., [13, 16]).
9. Interpreting this map, some of the changes are that Tos4, a putative transcription factor, is predicted to move from low expression in the control, to the nucleus under hydroxyurea treatment; Thr1, a homoserine kinase, is predicted to move from the cytoplasm to a nucleolar or punctuate localization, and Spc34, a spindle pole component, is predicted to move from a punctuate localization to low expression. Visual inspection of the images (Fig. 4b) confirms that the interpretation is largely correct, with great increases in the number of cells showing expression for Tos4 and Thr1 under hydroxyurea (HU). We note that the relocalization of Thr1 under HU treatment to this unusual pattern appears to be a new discovery.

3.7 Conclusions

We show that a deep learning pipeline can be used to visualize changes in cell morphology and subcellular localization. To our knowledge, this is still the only end-to-end, unsupervised approach for systematically identifying changes in protein localization directly in the feature space, so that phenotype/localization classes need not be defined. We believe the visualization methods presented here will allow researchers to more easily explore and validate the changes in cell phenotypes they observe in high-throughput microscopy screens.

4 Notes

1. Prices for deep-learning capable workstations are more affordable than you might think. At the time of writing, a system suitable for running the analysis described here (including SSD storage, basic Intel Xeon processor, and nvidia A2000 GPU) is less than $5000 USD at Dell.com.
2. Images from high-throughput microscopy screens are big data. Even the "toy" example here uses about 25GB of image data (including multi-channel images and segmentations masks). We suggest you convert your images to tif (or tiff) format. Most microscopes produce images in proprietary formats, so you will have to figure out how to convert these to tif or PNG (or some other format that can be opened by python's image analysis packages). Remember: don't compress your images when converting/exporting them. Even if the microscope software thinks this is a good idea.
3. Python is hard to install. You will probably have to use conda. With Windows 10, and later, it is now easy to run Ubuntu under windows. So, the suggested installation would be: Windows, Ubuntu, conda, python, tensorflow. Still hard.
4. You will have to figure out how to download and install codes from github. The simplest way is to select the "Code" button and then select "Download ZIP."
5. Any cell segmentation method that produces segmentation masks can be used here. We have tried a few different approaches over the years, and we found that as long as there is not too much mis-segmented noise, the next steps will usually work just fine.
6. For the applications here, we will work with crops that contain at least one cell. We won't worry if there are other cells in the picture; all the machine learning algorithms will work on the image pixels directly, and the different cells in the same image are assumed to be identical. Indeed, we simply average over all the crops to obtain a representation for each strain/group of

images. For applications where cell-to-cell variability is important, it is better to actually use the cell boundaries identified in the segmentation mask and block out the other cells in the crop.

7. We note that we have used paired-cell inpainting to learn features for our data in this example. However, what follows can be done with any image feature set. For example, these could be classical image analysis features (that can be computed with Cell Profiler [29]), interpretable, rationally designed features for yeast confocal microscopy [23], features extracted from a supervised deep learning method [16], or other self-supervised methods [8]. We have tried a few different feature sets over the years, and it seems like any features that can capture the phenotypes will work fine here.
8. One would think that simply averaging is not the best way to "pool" together the information over many cells in the image, and several others have been considered [12]. We have experimented over the years with other ways of summarizing the feature vectors and never found something that worked better and/or was more satisfying. We believe this is because our intuition about averaging in high-dimensional spaces is wrong: one would think that averaging would cancel out interesting single-cell variation, but because the space is very high-dimensional, it is unlikely that dimensions will happen to vary in opposite directions. Averaging may actually be keeping much of the interesting information. However, with the availability of attention layers and trainable pooling functions, this is clearly an area where further exploration is needed.
9. The magnitudes of the arrows in the UMAP visualization must be interpreted with caution. The mapping between the high-dimensional space and the 2D visualization is nonlinear, which means that smaller changes in the high-dimensional space could appear larger in the visualization. On the other hand, local distances in the UMAP are more reliable, so we can get a sense of the direction of change.

Acknowledgments

We thank members of the Moses lab for their input. The analysis reported here was performed on infrastructure obtained with grants to AMM from the Canadian Foundation for Innovation (CFI) AMM is supported by a Canada Research Chair.

References

1. Bagheri N, Carpenter AE, Lundberg E et al (2022) The new era of quantitative cell imaging – challenges and opportunities. Mol Cell 82:241–247
2. Meijering E (2020) A bird's-eye view of deep learning in bioimage analysis. Comput Struct Biotechnol J 18:2312. https://doi.org/10.1016/j.csbj.2020.08.003
3. Moen E, Bannon D, Kudo T et al (2019) Deep learning for cellular image analysis. Nat Methods 16:1233–1246. https://doi.org/10.1038/s41592-019-0403-1
4. Graham S, Vu QD, Raza SEA et al (2019) Hover-Net: simultaneous segmentation and classification of nuclei in multi-tissue histology images. Med Image Anal 58:101563. https://doi.org/10.1016/j.media.2019.101563
5. Lu AX, Zarin T, Hsu IS, Moses AM (2019) YeastSpotter: accurate and parameter-free web segmentation for microscopy images of yeast cells. Bioinformatics 35:4525–4527. https://doi.org/10.1093/bioinformatics/btz402
6. Hollandi R, Szkalisity A, Toth T et al (2020) nucleAIzer: a parameter-free deep learning framework for nucleus segmentation using image style transfer. Cell Syst 10:453–458.e6. https://doi.org/10.1016/j.cels.2020.04.003
7. Balestriero R, Ibrahim M, Sobal V et al (2023) A cookbook of self-supervised learning. arXiv. https://doi.org/10.48550/arXiv.2304.12210
8. Lu AX, Kraus OZ, Cooper S, Moses AM (2019) Learning unsupervised feature representations for single cell microscopy images with paired cell inpainting. PLoS Comput Biol 15:e1007348. https://doi.org/10.1371/journal.pcbi.1007348
9. Kobayashi H, Cheveralls KC, Leonetti MD, Royer LA (2022) Self-supervised deep learning encodes high-resolution features of protein subcellular localization. Nat Methods 19:995–1003. https://doi.org/10.1038/s41592-022-01541-z
10. Midtvedt B, Pineda J, Skärberg F et al (2022) Single-shot self-supervised object detection in microscopy. Nat Commun 13:7492. https://doi.org/10.1038/s41467-022-35004-y
11. Razdaibiedina A, Brechalov A, Friesen H (2023) et al, PIFiA: self-supervised approach for protein functional annotation from single-cell imaging data. bioRxiv. https://doi.org/10.1101/2023.02.24.529975
12. Caicedo JC, Cooper S, Heigwer F et al (2017) Data-analysis strategies for image-based cell profiling. Nat Methods 14:849–863. https://doi.org/10.1038/nmeth.4397
13. Chong YT, Koh JLY, Friesen H et al (2015) Yeast proteome dynamics from single cell imaging and automated analysis. Cell 161:1413–1424. https://doi.org/10.1016/j.cell.2015.04.051
14. Mattiazzi Usaj M, Sahin N, Friesen H et al (2020) Systematic genetics and single-cell imaging reveal widespread morphological pleiotropy and cell-to-cell variability. Mol Syst Biol 16:e9243. https://doi.org/10.15252/msb.20199243
15. Lu AX, Lu AX, Schormann W et al (2020) The Cells Out of Sample (COOS) dataset and benchmarks for measuring out-of-sample generalization of image classifiers. arXiv. https://doi.org/10.48550/arXiv.1906.07282
16. Kraus OZ, Grys BT, Ba J et al (2017) Automated analysis of high-content microscopy data with deep learning. Mol Syst Biol 13:924. https://doi.org/10.15252/msb.20177551
17. Cox MJ, Jaensch S, Van de Waeter J et al (2020) Tales of 1,008 small molecules: phenomic profiling through live-cell imaging in a panel of reporter cell lines. Sci Rep 10:13262. https://doi.org/10.1038/s41598-020-69354-8
18. Cuccarese MF, Earnshaw BA, Heiser K et al (2020) Functional immune mapping with deep-learning enabled phenomics applied to immunomodulatory and COVID-19 drug discovery. bioRxiv. https://doi.org/10.1101/2020.08.02.233064
19. Naik AW, Kangas JD, Sullivan DP, Murphy RF (2016) Active machine learning-driven experimentation to determine compound effects on protein patterns. elife 5:e10047. https://doi.org/10.7554/eLife.10047
20. Lu AX, Moses AM (2016) An unsupervised kNN method to systematically detect changes in protein localization in high-throughput microscopy images. PLoS One 11:e0158712. https://doi.org/10.1371/journal.pone.0158712
21. Lu AX, Chong YT, Hsu IS et al (2018) Integrating images from multiple microscopy screens reveals diverse patterns of change in the subcellular localization of proteins. elife 7:e31872. https://doi.org/10.7554/eLife.31872
22. Donovan-Maiye RM, Brown JM, Chan CK et al (2022) A deep generative model of 3D single-cell organization. PLoS Comput Biol 18:e1009155. https://doi.org/10.1371/journal.pcbi.1009155

23. Handfield L-F, Chong YT, Simmons J et al (2013) Unsupervised clustering of subcellular protein expression patterns in high-throughput microscopy images reveals protein complexes and functional relationships between proteins. PLoS Comput Biol 9:e1003085. https://doi.org/10.1371/journal.pcbi.1003085
24. Way GP, Kost-Alimova M, Shibue T et al (2021) Predicting cell health phenotypes using image-based morphology profiling. Mol Biol Cell 32:995–1005. https://doi.org/10.1091/mbc.E20-12-0784
25. La Manno G, Soldatov R, Zeisel A et al (2018) RNA velocity of single cells. Nature 560:494–498. https://doi.org/10.1038/s41586-018-0414-6
26. Koh JLY, Chong YT, Friesen H et al (2015) CYCLoPs: a comprehensive database constructed from automated analysis of protein abundance and subcellular localization patterns in Saccharomyces cerevisiae. G3 (Bethesda) 5:1223–1232. https://doi.org/10.1534/g3.115.017830
27. McInnes L, Healy J, Melville J (2020) UMAP: uniform manifold approximation and projection for dimension reduction. arXiv. https://doi.org/10.48550/arXiv.1802.03426
28. Huh W-K, Falvo JV, Gerke LC et al (2003) Global analysis of protein localization in budding yeast. Nature 425:686–691. https://doi.org/10.1038/nature02026
29. McQuin C, Goodman A, Chernyshev V et al (2018) CellProfiler 3.0: Next-generation image processing for biology. PLoS Biol 16:e2005970. https://doi.org/10.1371/journal.pbio.2005970

Chapter 16

Learning Morphological, Spatial, and Dynamic Models of Cellular Components

Huangqingbo Sun and Robert F. Murphy

Abstract

In this chapter, we describe protocols for using the CellOrganizer software on the Jupyter Notebook platform to analyze and model cell and organelle shape and spatial arrangement. CellOrganizer is an open-source system for using microscope images to learn statistical models of the structure of cell components and how those components are organized relative to each other. Such models capture the statistical variation in the organization of cellular components by jointly modeling the distributions of their number, shape, and spatial distributions. These models can be created for different cell types or conditions and compared to reflect differences in their spatial organizations. The models are also generative, in that they can be used to synthesize new cell instances reflecting what a model learned and to provide well-structured cell geometries that can be used for biochemical simulations.

Key words Machine learning, Image analysis, Image processing, Cell organization, Organelle modeling, Organelle shape

1 Introduction

Molecules located in various organelles and interacting with each other create specific chemical environments for cell functions [1–3]. There are dozens of major cellular organelles and thousands of other structures, many of which have a high degree of variation from cell to cell in their number and spatial distribution and from organelle to organelle in their size and shape. However, textbook example images of cell structures are typically hand-drawn cartoons or a single example image that does not reflect the morphological variance in the structure. Visual examination, or classic descriptive models that map the spatial patterns of molecules to probability vectors for a set of predefined categories, may fail to capture the intrinsic variance within the patterns. Cell biology is going through

Christoph Wülfing and Robert F. Murphy (eds.), *Imaging Cell Signaling*, Methods in Molecular Biology, vol. 2800,
https://doi.org/10.1007/978-1-0716-3834-7_16,

the transition from rough approximations of cell structure shapes and spatial distributions to spatially and morphologically accurate models of organelles.

Computational methods are expected to create automated, compact, statistically accurate generative models of cellular structure and organization from large collections of biological data [4–7]. CellOrganizer takes a collection of cell images with fluorescent tags for one or more proteins or organelles, automatically detects and parameterizes the major components in the images, and fits statistical models to characterize their morphological and spatial features [9]. For example, CellOrganizer can take fluorescence microscope images of many cells expressing a GFP-tagged lysosomal protein and construct a model of how the cell and nuclear shapes vary among those cells, how many lysosomes are typically found per cell, how they vary in size, and where they are typically located relative to the cell and nuclear boundaries. Cell and nuclear shape have been modeled using diffeomorphic models, principal component analysis, medial axis/ratio models, and spherical harmonic transforms [10–14]. Vesicular organelles (e.g., mitochondria) have been modeled using Gaussian mixture models and spherical harmonic transforms [10, 14, 15]. Tubular organelles have been modeled by growth-based network models [16, 17].

CellOrganizer is primarily a MATLAB package, but it has recently been made available through a Docker container containing compiled executables that do not require a MATLAB license. The Docker container also provides a Jupyter Notebook platform that provides lightweight and convenient access for the broader bioimage informatics community. CellOrganizer for Jupyter Notebook is accessed by a small number of interface functions for constructing models from input images, comparing models, and synthesizing new instances from learned models (*see* **Note 1**). Here, we provide protocols for installing CellOrganizer for Jupyter Notebook, organizing your own images for use with CellOrganizer, constructing spatial and shape models, comparing different models, and synthesizing new cell images and movies from CellOrganizer models.

2 Materials

CellOrganizer for Jupyter Notebook runs in a Docker container (*see* **Note 2**). Jupyter Notebook is a web-based interactive platform that creates, edits, and runs notebook with a mix of computer code, formatted text, and computed results or graphics. It run inside the CellOrganizer Docker container.

3 Methods

3.1 Installing Docker

Prior to using CellOrganizer for Jupyter Notebook, install the Docker software following the instructions for Docker installation at https://docs.docker.com/get-docker/. Docker is available for macOS (Sierra or newer), Linux (Debian 8, 9, 10, and CentOS 7), and Windows platforms.

3.2 CellOrganizer for Jupyter Notebook Installation

1. For Linux users, please refer to https://docs.docker.com/engine/install/linux-postinstall/ for post-installation procedures.

3.2.1 Starting Docker

2. In Linux, the Docker daemon starts when the machine is booted, but in MacOS and Windows, it must be manually started (or included in startup items). Find the Docker Desktop application and open it.

3.2.2 Pulling the CellOrganizer Docker Container

1. We will use a terminal/Command Prompt window to issue docker commands. In MacOS, open the */Applications/Utilities* folder, then double-click Terminal. In Windows, press the Windows key + R, type `cmd` or `cmd.exe` in the Run command box, and press Enter.
2. If you have an older version of the CellOrganizer Docker container, you can remove it by issuing these commands in the Terminal/Command Prompt before pulling the new one:
   - `docker stop $(docker ps -aq)`
   - `docker rm $(docker ps -aq)`
   - `docker rmi -f $(docker images -q)`
3. Pull the Docker image of CellOrganizer from DockerHub:
   - `docker pull murphylab/cellorganizer-jupyter:latest`

   You can verify that the container was pulled by typing
   - `docker images`

3.2.3 Starting CellOrganizer

1. Create a folder that CellOrganizer can read from and write to in order to run notebooks and save results. In the examples below, creating a folder "CellOrganizerDocker" inside the Desktop folder is assumed. The name of the folder can be changed but the commands below must be changed to match.
2. Depending on your operating systems and settings, you may need to explicitly make the folder you created available to Docker. To do so, click on *settings* in the Docker Desktop app, click on *Resources* > *File Sharing*, and select the folder you created by clicking (+) and searching for the folder you created in your desktop, then click Apply & Restart (Fig. 1c).

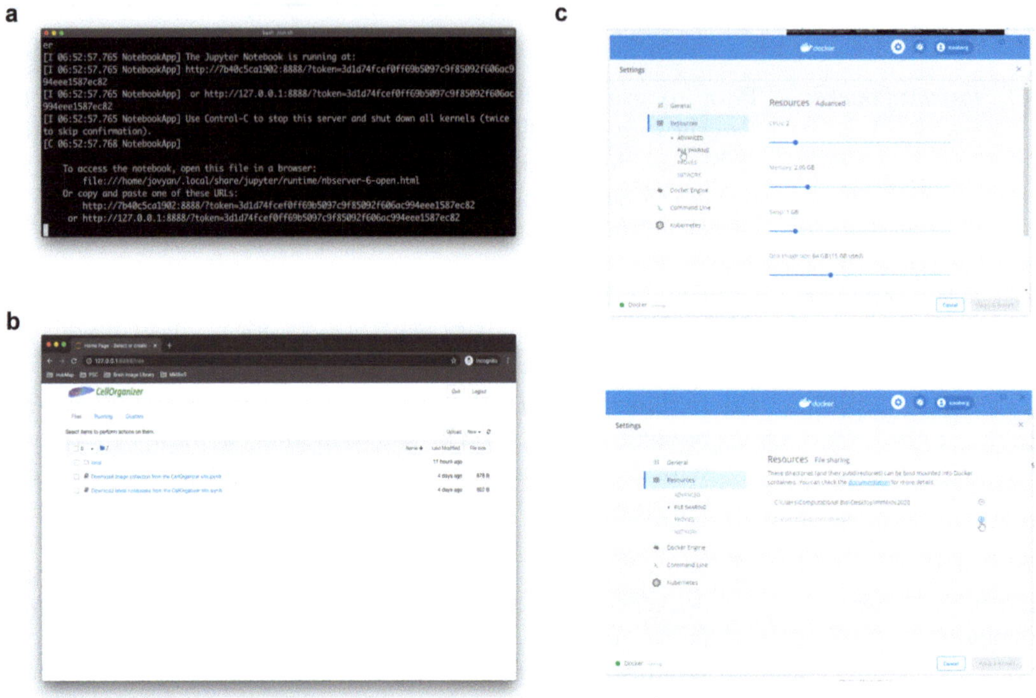

Fig. 1 (**a**) terminal window after successful launching the Docker container for CellOrganizer. (**b**) the main Jupyter Notebook page for CellOrganizer displayed in a web browser. (**c**) Illustration of changing Docker settings

3. To start the container, type the appropriate following command,

   in Linux:

   - `chmod a+rwx -R ~/Desktop/CellOrganizerDocker/ && docker run --rm -p 8888:8888 -v ~/Desktop/CellOrganizerDocker:/local --memory="4g" --cpus=2 -e JUPYTER_LAB_ENABLE=yes murphylab/cellorganizer-jupyter:latest`

   in macOS:

   - `docker run --rm -p 8888:8888 -v ~/Desktop/CellOrganizerDocker:/local --memory="4g" --cpus=2 -e JUPYTER_LAB_ENABLE=yes murphylab/cellorganizer-jupyter:1`

   in Windows:

   - `docker run --rm -p 8888:8888 -v "%userprofile%"\Desktop\CellOrganizerDocker:/local --memory="4g" --cpus=2 -e JUPYTER_LAB_ENABLE=yes murphylab/cellorganizer-jupyter:latest`

4. If the container started properly, you should see a screen similar to Fig. 1a. The URL at the end of the message will be different each time, as the URL contains a token. Copy the URL, paste it in a web browser and press enter. You should see a page like that shown in Fig. 1b.
5. The folder you created will be connected to the *local* folder on the main Jupyter Notebook webpage.
6. Example CellOrganizer Notebooks, models, and images are stored on online servers for use in tutorials. On the main page, click on the *Download_files* notebook to run it; it will pull notebooks, models, and images needed for the demos and tutorial modules and place them in folders in the *local* folder (*see* **Note 3**).
7. Download the ZIP file for the github repository for this chapter from https://github.com/sun-huangqingbo/CellOrganizer_Notebooks/archive/refs/heads/main.zip onto your computer and unzip it if it does not automatically unzip. It will create a folder called *CellOrganizer_Notebooks-main*. Move that folder to the *notebooks* folder that has been created inside the main "CellOrganizerDocker" folder.
8. On the Jupyter Notebook main page, click on the *local* folder, then on the *notebooks* folder, then on the *CellOrganizer_Notebooks-main* folder. The notebooks to be used in the following sections are contained there.

3.3 Preparing Files for Use with CellOrganizer

CellOrganizer requires images containing single cells. It also needs to know which fluorescence channels should be used to construct which parts of the model. There are three major options for how to do this (*see* **steps 4a–c**). The first is to have a separate file for each cell. The second is to have multiple cells in each image (the way they were presumably acquired) and provide an additional file containing a "cell mask" to specify where a cell is located within the image; this file is a binary image in which the "on" pixels are to be included in the cell. The third is to have multiple cells in an OME-TIFF file and include Regions of Interest (ROIs) for each cell.

1. Open the Notebook of Module 1A in the Jupyter Notebook platform (*see* **Note 4**).
2. Run *Code Cell* 1 to set up general options for CellOrganizer. See Table 1 for basic CellOrganizer options (*see* **Note 5**).
3. Run *Code Cell* 2 to create Python *lists* for storing image paths later.
4. Choose one of the following alternatives.

Table 1
Basic CellOrganizer options

Options	Descriptors	Example settings
model.name	Holds the name of the model	"module_1"
train.flag	Selects what model is going to be trained	"nuclear," "cell," "framework," or "all"
verbose	Selects to display extended information	"true" or "false"
downsampling	The downsampling vector to be used during preprocessing	[2, 8]
model.resolution	Holds the information of the dimensionality of the images	[0.05, 0.05, 0.2]
if_skip_cell_nuclear_model	Boolean condition to skip building a nuclear model	"true" or "false"
min_obj_size	Threshold value for determining whether the object should be saved	20
max_obj_size	Threshold value for determining whether the object should be saved	400
cell.class	Holds the cell membrane model class	"cell_membrane"
nucleus.class	Holds the nuclear membrane model class	"nuclear_membrane"
protein.class	Holds the protein model class	"vesicle"
cell.type	Holds the cell membrane model type	"spharm_rpdm"
nucleus.type	Holds the nuclear membrane model type	"spharm_rpdm"
protein.type	Holds the protein model type	"spharm_obj"
model.spharm_rpdm. postprocess	This specifies whether alignment and size normalization, should be done after parameterization	"true" or "false"
model.spharm_rpdm. maxDeg	This specifies the degree up to which spherical harmonics should be calculated. Valid values are positive integers	31
model.spharm_rpdm. latent_dim	This specifies how many latent dimensions should be used for modeling the shape space. Valid values are positive integers	15
hd_thresh	Threshold for error tolerance for a given cell. If above this parameter the cell is discarded	10

4a. Prepare separate image files for each cell. Replace the file paths in *Code Cell* 3 with your own paths and run the Cell to append image file paths of each cell and channel into the lists created earlier.

4b. Prepare a single image containing multiple cells and an additional cell mask to specify where a cell is located within the image. Replace the file paths in *Code Cell* 4 with your own paths and run the Cell to append image file paths of each channel into the lists and store the path of the cell mask.

4c. Prepare an OME-TIFF file and ROIs to for each cell. To do this, open the Notebook of Module 1B in the Jupyter Notebook platform. Run *Code Cell* 1 to import required packages and define required image processing Python functions. Replace the directory paths in *Code Cell* 2 and run it. Write OME-TFF image files by running *Code Cell* 3. Now, return to the Notebook of Module 1A. Replace the file paths in *Code Cell* 5 with the paths of OME-TIFF images just created and run the Cell to append cell and DNA image file paths following their associated ROIs into the lists created earlier.

3.4 Building Generative Shape Models for Cells and Nuclei and Generating Single-Cell Images

1. Open the Notebook of Module 2A. This notebook contains example code that can be run to create shape models using `img2slml` and then illustrates some things that can be done with those models using `slml2info`, `slml2report`, and `slml2img` (*see* **Note 6**).
2. The spherical harmonic transform provides a representation of a shape by mapping the curvature map of a shape onto an orthogonal space spanned by the spherical harmonic basis. In CellOrganizer, cell and cell nucleic shapes are parameterized by spherical harmonic descriptors and referred to *SPHARM-RPDM* models (Fig. 2). Run *Code Cell* 1 to set up appropriate parameters for building *SPHARM-RPDM* models. The inputs to *SPHARM-RPDM* models are expected to be 3D images containing single cells as prepared in Subheading 3.2.
3. Replace the path of the images and directory in *Code Cell* 2 with your own paths and run the *Code Cell* to define the input files. Ideally, put all files to be used to build a model into a single directory (but all files for a single channel could be in a subdirectory). The idea is to allow wildcards to be used easily (*see* **Note** 7).
4. Run *Code Cell* 3 to build a *SPHARM-RPDM* model of cell and cell nuclear shape using the provided images.
5. Like **step 3**, replace the path of the directory in *Code Cell* 4 with a directory containing own images and run the *Cell* to define another group of input files.
6. Run *Code Cell* 5 to build another *SPHARM-RPDM* model using images provided in **step 5**.
7. Open the Notebook of Module 2B. This notebook generates reports on the models just built.
8. Run *Code Cell* 1 to load the first *SPHARM-RPDM* model and generate a report of it.

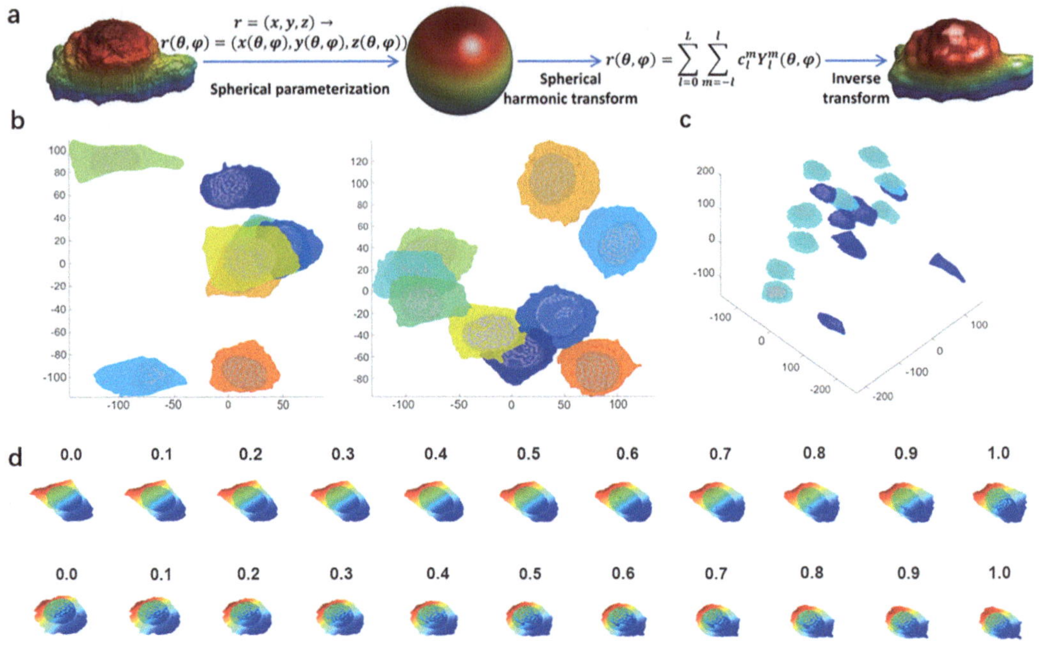

Fig. 2 (**a**) Illustration of Shape modeling using spherical Harmonic transforms. In the first step, the 3D surface (represented as a mesh) is mapped onto a unit sphere. Then, the spherical harmonic transform is performed to get coefficients to represent the surface. Those coefficients can be used to reconstruct the original shape. (Figure reproduced from Fig. 1 in reference [13]); (**b**) two cell shape space plots generated by the two SPHRM-RPDM models built in Subheading 3.3. After computing the spherical harmonic descriptors, their dimensionality is reduced to 2 by principal component analysis. A 2D projection of each cell is shown centered at the values of its principal components; (**c**) a combined view of the cells in panel **b** to illustrate comparison of the shapes of two cell populations; (**d**) two example trajectories of shape evolution illustrating the morphing of one cell shape to another. The numbers above each panel show the normalized distance to the source

9. Sequentially run *Code Cells* 2–4 to see the cell shape space generated by the first *SPHARM-RPDM* model and example trajectories of cell shape evolutions.
10. Run *Code Cell* 5 to generate a report of the second model.
11. Like **step 8**, generate figures showing the cell shape space and trajectories generated by the second *SPHARM-RPDM* model.
12. Next, we compare two cell shape models. Run *Code Cell* 8 to generate a shape model comparison report.
13. Run *Code Cell* 9 to see the combined cell shape space of the two *SPHARM-RPDM* models built earlier.
14. Open the Notebook of Module 2C. This Notebook creates a synthetic cell using one of the models just built.
15. Run *Code Cells* 1 and 2 to create a synthetic cell and generate its images. The notebook creates a directory with the same name as the model that will contain a directory named *img*

containing a folder named *cell1*. This will contain files *cell1.tif* and *nucleus.tif* which will contain 2D images corresponding to consecutive slices of the 3D cell.

16. Run *Code Cells* 3 and 4 to show the synthetic cell images.
17. To generate a movie of the evolution of one cell shape to another using one of the built models, run *Code Cell* 5. Then, run *Code Cell* 6 to visualize it.

3.5 Building Shape Models for Organelles

1. Open the Notebook of Module 3A. This Notebook generates an organelle *spharm-obj* model and plots organelle spatial distribution. *Spharm-obj* models provide a highly detailed representation of the size, shape, and subcellular distribution of organelles and other subcellular structures. They use the same SPHARM-RPDM capability that is used for modeling cell and nuclear shape.
2. Replace the image directory path in *Code Cell* 1 with your own image paths, the images are supposed to be 2D or 3D microscopy images as prepared in Subheading 3.2. Edit parameters of the model as desired: the resolution of images and the model, number of images to use, size thresholds of organelle objects, model filename, etc. Then, run this *Code Cell* to generate a shape model.
3. Run *Code Cell* 2 to generate a report of the model just built.
4. Run *Code Cells* 3 and 4 to see the shape space associated with the shape model (like the example in Fig. 3a).
5. Run *Code Cells* 5 and 6 to see the spatial distribution of the organelle (Fig. 3b).
6. Notebook of Module 3B uses `slml2report` to compare *spharm-obj* models of two different organelles. The comparison includes comparing the organelle shapes and their subcellular distributions. We provide our trained shape models along with this Notebook for convenience in illustration. Open the Notebook of Module 3B and run *Code Cell* 1 to import the packages needed.
7. Run *Code Cell* 2 to generate a comparison between models of Golgi apparatus and nucleoli.
8. Run *Code Cells* 3–5 to see the differences in their shapes and spatial distributions.
9. Like **step** 7, run *Code Cell* 6 to generate a comparison between models of Golgi apparatus and desmosome.
10. Again, run *Code Cells* 7–9 to see the differences in their shapes and spatial distributions.

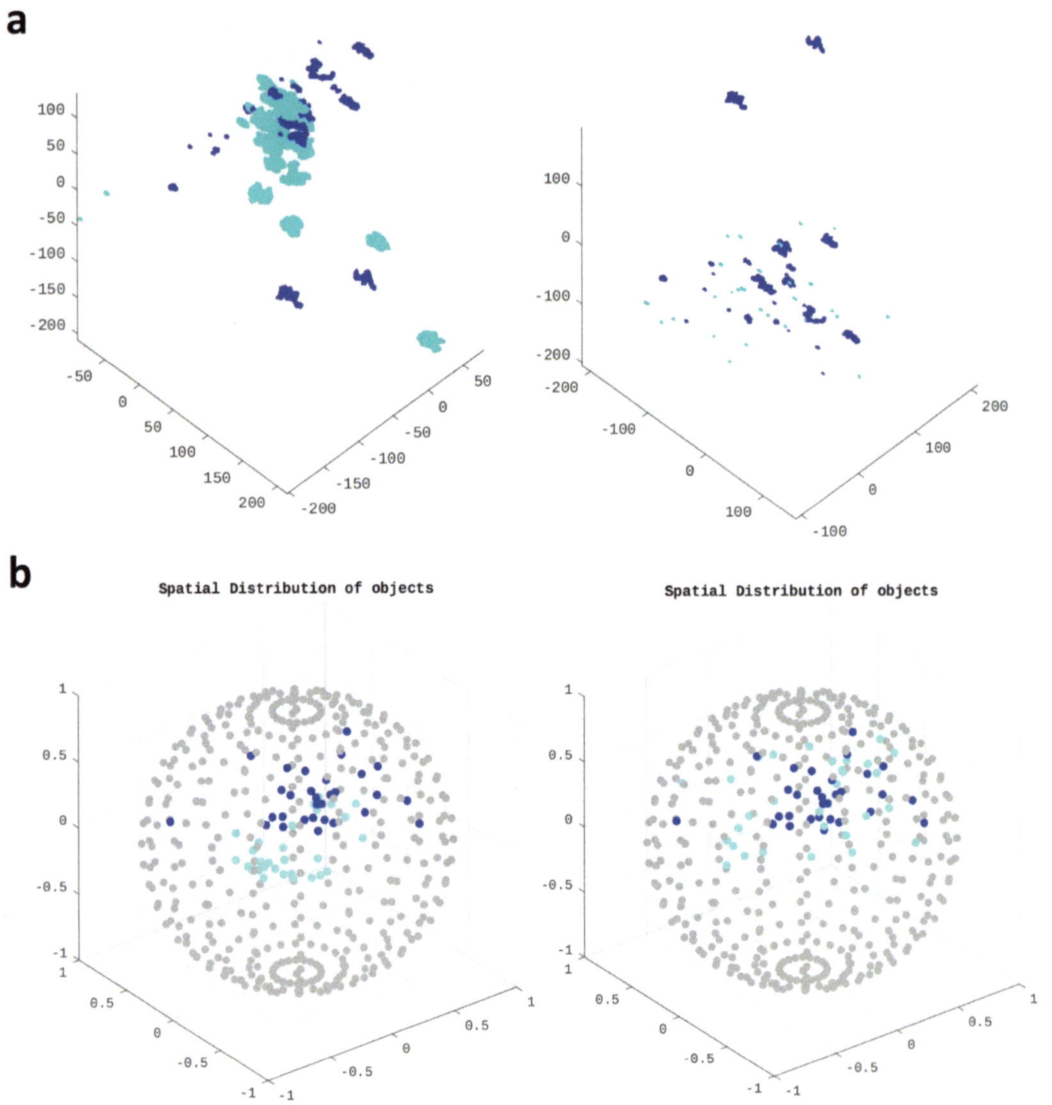

Fig. 3 (**a**) example shape spaces of two organelle patterns. Left: nucleoli (cyan) and Golgi apparatus (blue); right: desmosome (cyan) and Golgi apparatus (blue). The shape spaces are generated by *spharm-obj* models using reduced spherical harmonic descriptors; (**b**) spatial distributions of organelles for the same organelles as in panel a. The cell surface is mapped onto a unit ball (shown in gray dots), and the centroids of organelles are shown in blue and cyan dots

3.6 Parameterizing Object Shape Representations in CellOrganizer

1. Open the Notebook of Module 4. This Notebook illustrates how to directly generate a shape descriptor using spherical harmonics for a provided shape without having to construct a full CellOrganizer model and how to create a visualization of continuous shape changes by interpolating between two shapes (*see* **Note 8** and Fig. 4).
2. Run *Code Cell* 1 to import required packages in Python (*see* **Note 9**).

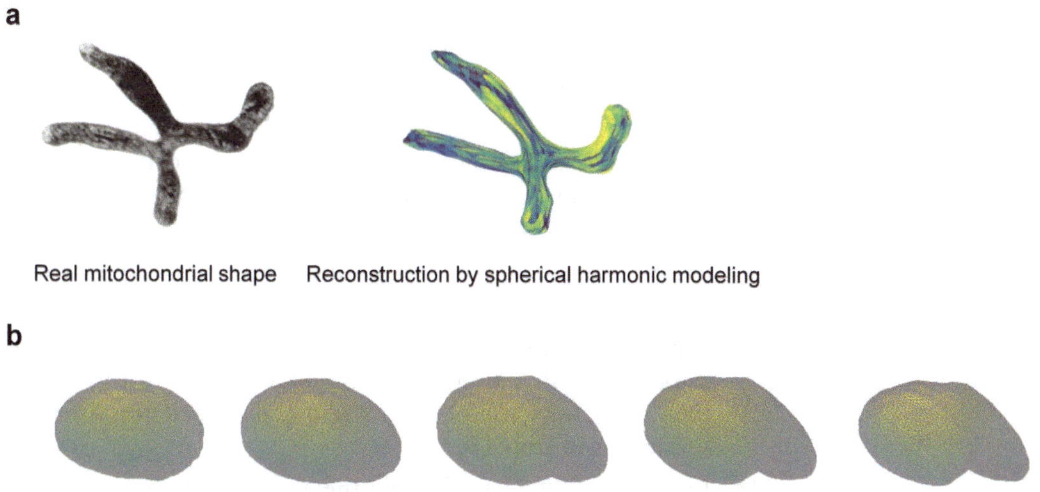

Fig. 4 (**a**) an example of modeling an individual mitochondria shape observed from electron microscopy using spherical harmonic transform; (**b**) an example trajectory of morphing one mitochondrion with an ellipsoidal shape into a mitochondrion with a curved shape

3. Replace the directory and image paths in *Code Cell* 2 and run *Code Cell* 2 to read the image. Here, the input image is expected to be a binarized image containing the shape of a single object with 0 and 255 indicating "off" and "on" pixels.
4. Run *Code Cell* 3 to compute the spherical harmonic descriptors associated with the input object image.
5. Run *Code Cell* 4 to compute the reconstructed object surface from computed spherical harmonic descriptors and run *Code Cell* 5 to visualize it.
6. Run *Code Cell* 6 to reconstruct the object from spherical harmonic descriptors by sampling pixels/voxels inside the reconstructed object surface. Then run *Code Cells* 7 and 8 to visualize the reconstructed object (see example in Fig. 4a).
7. Now, we are going to create a "continuous" dynamic of morphing one object shape to another by interpolating between their spherical harmonic descriptors. Run *Code Cell* 9 to define plotting and linear interpolation functions needed later.
8. First, we try to directly interpolate between two spherical harmonic descriptors associated with two valid objects. Replace the image paths in *Code Cell* 10 with your image paths and read in two object images. The images are expected to be in the same format as **step 3**.
9. Run *Code Cell* 11 to compute their spherical harmonic descriptors.

10. Run *Code Cells* 12 and 13 to visualize the reconstructed surfaces associated with the spherical harmonic descriptors computed in **step 9**.
11. Run *Code Cell* 14 to flatten the spherical harmonic descriptors and run *Code Cell* 15 to see the process of shape morphing (see Fig. 4b for an example). Run *Code Cell* 16 to save it as a video.
12. Next, we try a different way to create the shape morphing by interpolating in a lower dimensional space. Run *Code Cell* 17 to read in a valid CellOrganizer shape model.
13. Run Code Cell 18 to select spherical harmonic descriptors associated with the first and second shapes from this model, then run *Code Cells* 19 and 20 to visualize their reconstructed surfaces as in **step 10**.
14. Run *Code Cell* 21 to reshape the array of spherical harmonic descriptors associated with all shapes in this model and perform PCA for dimensionality reduction. Run *Code Cells* 22 and 23 to see the reconstructed surfaces using the dimensional-reduced descriptors as opposed to the reconstruction with full descriptors in **step 13**
15. Run *Code Cells* 24 and 25 to linearly interpolate the dimensional-reduced spherical harmonic descriptors and visualize the resulting dynamics.

4 Notes

1. There are four main CellOrganizer functions. `img2slml` makes models from images. `slml2img` synthesizes images from models. `slml2info` describes a model. slml2report compares two different models. There are also auxiliary functions for working directly with object shape representations. Comprehensive documentation for CellOrganizer can be found at http://www.cellorganizer.org/documentation/cellorganizer/. The Python versions of these functions call compiled executables created from the CellOrganizer Matlab code.
2. A Docker image is a standalone, executable file, composed of multiple layers, containing all the files, libraries, and dependencies for a Docker container to run the software. Docker images create containers at runtime. A container is a runtime environment that packages code and all its dependencies.
3. When you run the CellOrganizer Docker container, it connects to a local folder that will appear as */local* in the main Jupyter window. After you have run the *Download_files* Notebook (during installation), three folders will have been created within this local folder: *models*, images, and *notebooks*. Output files from most notebooks are created in a *results* folder (which will automatically be created by the first notebook that uses it).

4. Each Jupyter Notebook is run in a separate *process* on your computer. After finishing using a notebook, you can stop that process by going to the *Running* tab and selecting that notebook. To stop the main Jupyter Notebook server, you can click on Quit at the top of the main Notebook window or go to Terminal and click Control+c. Since the containers have been saving your results to the local folder you created at the beginning, it means that stopping the Jupyter notebook does not erase the results you have computed so far (but anything that has not been saved, including any unsaved changes to the notebook itself, will be lost.
5. When running these notebooks, the output from Python appears below each Code cell in the Notebook as it is run, while the output from the compiled CellOrganizer MATLAB executables appears in the terminal window that was used to start the Docker container. MATLAB errors listed in the terminal window will include a trace of the MATLAB module and line numbers that generated the error. For those interested in debugging a problem, the line numbers can be referenced against the MATLAB code, which can be found at https://github.com/murphygroup/cellorganizer.
6. For the sake of reducing computing time, we use a small number of images in the Notebook examples; this can be changed by editing the associated numbers in the code.
7. When organizing your own images for use with CellOrganizer, letting CellOrganizer know what files to use to construct a model is done using wildcards (e.g., "*Expt1Sample*Nuclear.tif*"), and it is helpful to organize your files to facilitate this.
8. There are three main functions used in Module 4: `Image2SPHARMparameterization` takes an image and converts it into a structure that contains the spherical harmonic parameters along with additional metadata information; `SPHARMparameterization2Mesh` takes a generated structure of spherical harmonic parameters and converts it into a mesh structure that contains vertices and faces; `SPHARMparameterization2Image` takes a generated structure of spherical harmonic parameters and converts it into a 3D array or image.
9. Inside the container, the CellOrganizer python library can be found at /opt/conda/lib/python3.7/site-packages/cellorganizer.

Acknowledgments

The original research upon which these protocols are based was supported by National Institutes of Health grants R01 GM090033 and P41 GM103712.

References

1. Mochly-Rosen D (1995) Localization of protein kinases by anchoring proteins: a theme in signal transduction. Science 268:247–251
2. Huh WK, Falvo JV, Gerke LC et al (2003) Global analysis of protein localization in budding yeast. Nature 425:686–691
3. Hung MC, Link W (2011) Protein localization in disease and therapy. J Cell Sci 124:3381–3392
4. Neumann B, Walter T, Hériché JK et al (2010) Phenotypic profiling of the human genome by time-lapse microscopy reveals cell division genes. Nature 464:721–727
5. Eliceiri KW, Berthold MR, Goldberg IG et al (2012) Biological imaging software tools. Nat Methods 9:697–710
6. Carpenter AE, Jones TR, Lamprecht MR et al (2006) CellProfiler: image analysis software for identifying and quantifying cell phenotypes. Genome Biol 7:1–11
7. Glory E, Murphy RF (2007) Automated subcellular location determination and high-throughput microscopy. Dev Cell 12:7–16
8. Hu Y, Osuna-Highley E, Hua J et al (2010) Automated analysis of protein subcellular location in time series images. Bioinformatics 26: 1630–1636
9. Murphy RF (2016) Building cell models and simulations from microscope images. Methods 96:33–39
10. Peng T, Murphy RF (2011) Image-derived, three-dimensional generative models of cellular organization. Cytometry A 79:383–391
11. Johnson GR, Buck TE, Sullivan DP et al (2015) Joint modeling of cell and nuclear shape variation. Mol Biol Cell 26:4046–4056
12. Rohde GK, Ribeiro AJ, Dahl KN, Murphy RF (2008) Deformation-based nuclear morphometry: capturing nuclear shape variation in HeLa cells. Cytometry A 73:341–350
13. Ruan X, Murphy RF (2019) Evaluation of methods for generative modeling of cell and nuclear shape. Bioinformatics 35:2475–2485
14. Zhao T, Murphy RF (2007) Automated learning of generative models for subcellular location: building blocks for systems biology. Cytometry A 71:978–990
15. Sun H, Fu X, Abraham S et al (2022) Improving and evaluating deep learning models of cellular organization. Bioinformatics 38: 5299–5306
16. Shariff A, Murphy RF, Rohde GK (2011) Automated estimation of microtubule model parameters from 3-D live cell microscopy images. IEEE Int Symp Biomed Imaging 2011:1330–1333
17. Shariff A, Murphy RF, Rohde GK (2010) A generative model of microtubule distributions, and indirect estimation of its parameters from fluorescence microscopy images. Cytometry A 77:457–466

Index

Christoph Wülfing and Robert F. Murphy (eds.), *Imaging Cell Signaling*, Methods in Molecular Biology, vol. 2800, https://doi.org/10.1007/978-1-0716-3834-7,

M

O

P

R

S

T

MIX
Papier aus verantwortungsvollen Quellen
Paper from responsible sources
FSC® C105338

If you have any concerns about our products,
you can contact us on
ProductSafety@springernature.com

In case Publisher is established outside the EU,
the EU authorized representative is:
Springer Nature Customer Service Center GmbH
Europaplatz 3, 69115 Heidelberg, Germany

Printed by Libri Plureos GmbH
in Hamburg, Germany